Taming the Truffle

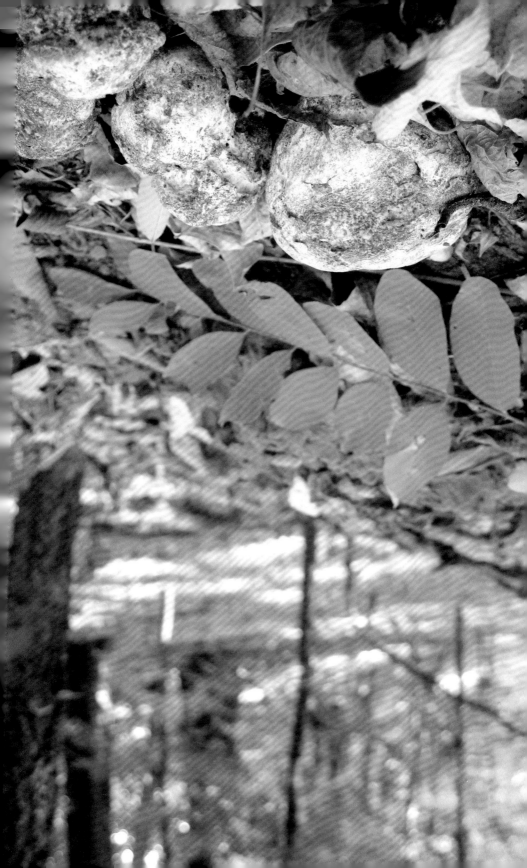

TAMING
THE
TRUFFLE

The History, Lore, and Science
of the Ultimate Mushroom

IAN R. HALL
GORDON T. BROWN
ALESSANDRA ZAMBONELLI

TIMBER PRESS

Frontispiece, page 2: in the woods just outside Budapest, Hungary. The truffles are *Mattirolomyces terfezioides*. Photograph by I. R. Hall. Opposite: Hunting for truffles, Italy. Photograph by I. R. Hall.

Note: For a wide variety of reasons, many truffle plantations never produce. Consequently, investing in the cultivation of truffles and other ectomycorrhizal mushrooms must still be regarded as a speculative venture and potentially high risk. This book was prepared using the best information available at the time it was written, and all due care was exercised in its preparation. Sources of additional information on mycorrhizas, truffles, and ectomycorrhizal mushrooms and Web site addresses can be found in the list of references. However, any subsequent action that relies on the accuracy of the information contained in this book is the sole commercial decision of the user and is taken at his or her own risk. Accordingly, the authors and the publisher disclaim any liability whatsoever in respect of any losses or damages arising out of the use of this information.

Small parts of this book have their origins in *The Black Truffle: Its History, Uses and Cultivation* (1994, New Zealand Institute for Crop and Food Research Ltd., Christchurch) by Ian Hall, Gordon T. Brown, and James Byars.

Copyright © 2007 by Ian R. Hall, Gordon T. Brown, and Alessandra Zambonelli. All rights reserved.

Published in 2007 by
Timber Press, Inc.
The Haseltine Building
133 S.W. Second Avenue, Suite 450
Portland, Oregon 97204-3527, U.S.A.
www.timberpress.com

For contact information regarding editorial, marketing, sales, and distribution in the United Kingdom, see www.timberpress.co.uk.

Designed by Susan Applegate
Printed in China

Library of Congress Cataloging-in-Publication Data

Hall, Ian R. (Ian Robert), 1946–
    Taming the truffle: the history, lore, and science of the ultimate mushroom/Ian R. Hall, Gordon T. Brown, and Alessandra Zambonelli.
        p. cm.
    Includes bibliographical references and index.
    ISBN-13: 978-0-88192-860-0
    1. Truffles. 2. Truffle culture. 3. Truffières. I. Brown, Gordon T. (Gordon Thomas), 1950– II. Zambonelli, Alessandra. III. Title.
    SB353.5.TT78H344 2008
    635'.8—dc22   2007019017

A catalog record for this book is also available from the British Library.

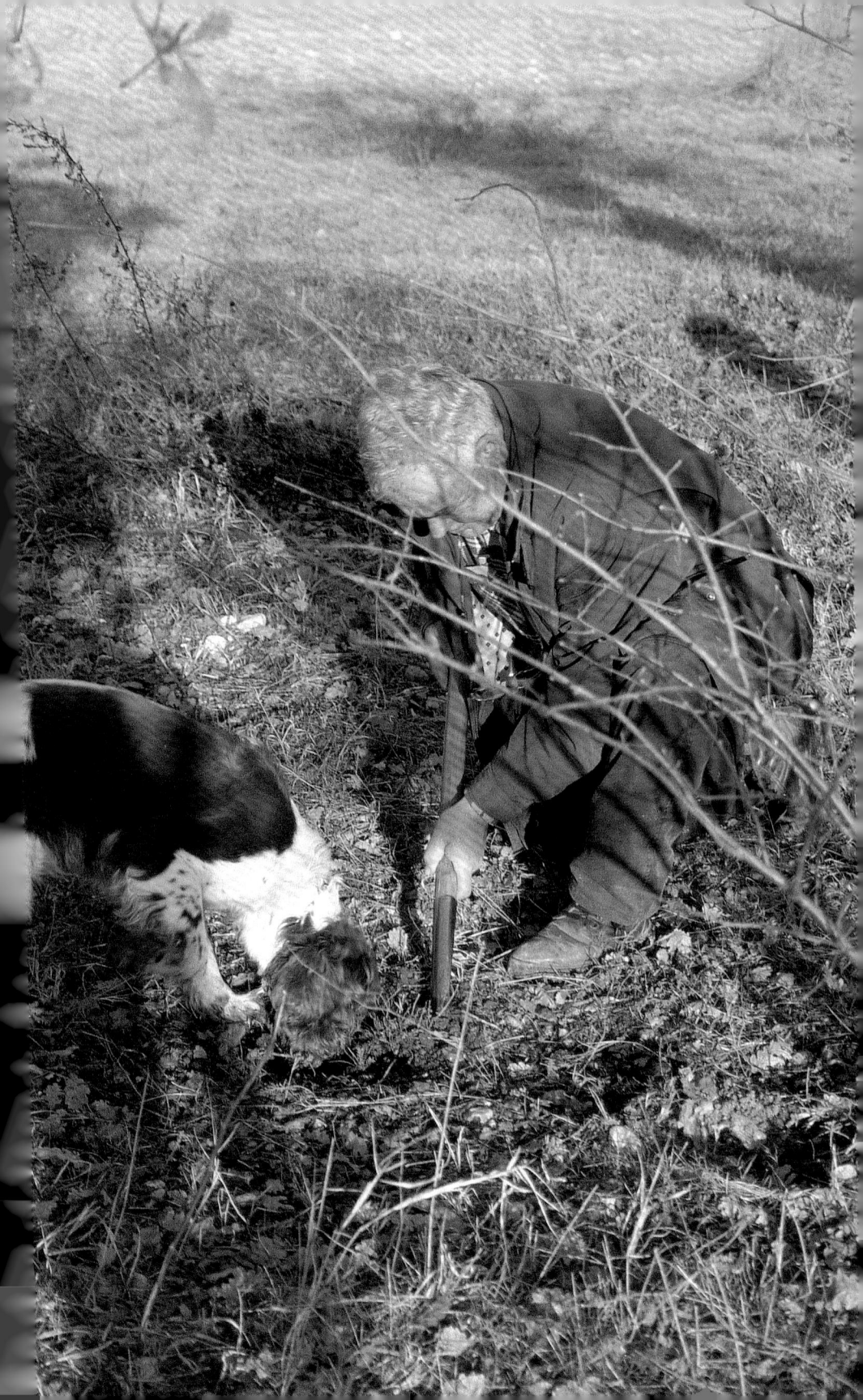

CONTENTS

9 Preface

11 Acknowledgments

15 ONE From the Past Comes the Present

Regions of Renown • *La Grande Mystique* • The Golden Age of Truffles • Magical Powers and Medicinal Virtues

34 TWO Science to the Rescue

Mycorrhiza: An Intimate Relationship • Types of Mycorrhizas • Structure of Ectomycorrhizas • Underground Warfare • The Collapse of Truffle Production • New Techniques

53 THREE Identifying Truffle Species

A Quick Lesson in Classification, Nomenclature, and Taxonomy • Identifying Species of Truffles • Périgord Black Truffle • Burgundy or Summer Truffle • Winter Truffle • Bagnoli Truffle or *Truffe Mésentérique* • Smooth Black Truffle • Asiatic Truffles • Italian White Truffle • Bianchetto and Other Pale-Coloured European Species • North American Pale-Coloured Truffles • Oregon Black Truffles • Sweet Truffle • Desert Truffles • Inedible and Poisonous Truffles • Other Species of Truffle • False Truffles • Identifying Truffle Species from Their Mycorrhizas • Identifying Truffles Using Molecular Techniques

98 FOUR The Habitats of Some Commercial Truffles

Périgord Black Truffle • Burgundy or Summer Truffle • Italian White Truffle • Bianchetto • Desert Truffles

121   FIVE Establishing a Truffière

Competing Ectomycorrhizal Fungi ✦ Choosing a Site for a
Truffière ✦ Soil Texture and Drainage Characteristics ✦ Soil
Nutrients ✦ Site Preparation ✦ Planting Density and Tree
Arrangement ✦ Irrigation System ✦ Competing Ectomycorrhizal
Fungi on Nursery Stock ✦ Other Plant Specifications ✦ Choice
of Host Plant ✦ Double Cropping and Interplanting ✦ When and
How to Plant ✦ Desert Truffles

163   SIX Maintenance of Truffières

Maintenance Methods ✦ Working the Ground ✦ Brûlé ✦
Herbicides Versus Soil Cultivation ✦ Soil Moisture ✦ Fertilizers ✦
Nutrient Deficiency ✦ Plant Diseases ✦ Invertebrate Pests ✦ Truffle
Pests, Diseases, and Browsers ✦ Pruning ✦ Thinning and the
Rejuvenation of Truffières

198   SEVEN The Rewards

Le Cavage: The Harvest ✦ Regulations and Reality ✦ A Time-
Honoured Ritual ✦ Failure to Produce ✦ Cultivation Outside of
Europe ✦ Yields ✦ Prices ✦ Marketing by Fair Means or Foul ✦
Truffles Without Plants? ✦ The Future

  Appendix

231   1. Plants that form arbuscular mycorrhizas

233   2. Plants that may harbour fungi that might compete
    with truffle fungi

234 3. Some host plants of commercially important species of truffle

239 4. Combined production of Périgord black and winter truffles for France and Spain, 1990–2005

240 5. Total truffle production in Italy, 1950–2006

241 6. Some desert truffles

242 7. Climatic data for various centres adjacent to or with similar climates to truffle-producing areas

246 8. Physical and chemical properties of *Tuber magnatum* soils

247 9. Climatic features in Gard, Drôme, and Vaucluse and production figures for Carpentras

248 10. Factors that might trigger fruiting or affect productivity of truffières

249 11. Manganese fertilizers

249 12. Commonly used boron fertilizers

250 13. Authorized harvesting dates for truffles in France and Italy

251 Truffle Organizations and Plant and Equipment Suppliers

254 References

297 Index

PREFACE

ON SEVERAL OCCASIONS I have been asked, "What got you into all this?" My story began in the late 1970s, when I attended a conference in Colorado. During a lunch break I was seated at a table next to a group of Frenchmen. My year 6 French teacher, Mr. Townshend, would certainly testify that I had less than a schoolboy's knowledge of French, but I understood enough to overhear that the first "artificial" truffles had been harvested in France. From there developed the idea of growing truffles in New Zealand for out-of-season Northern Hemisphere markets. From that point on, the only thing that needed to be done was to simply collect or develop the necessary techniques. There was a delay of a few years while I tried to convince the powers that be that growing truffles in New Zealand was a better use of my time than upgrading upland native pastures to carry more sheep. By 1985, however, I had the green light, and with a grant from the Miss E. L. Hellaby Trust I began to piece together the puzzle.

During the next decade or so I was to make some excellent friends in the world of truffles, two of whom coauthor this book. I was also assisted by some very capable and hard-working colleagues. What kept me going were a couple of sentences from a novel by the physicist Fred Hoyle: "If you know something to be possible, it is much easier to find it yourself. It is when you mistakenly think something to be impossible that it gets really hard." Comments from the great Kiwi knocking machine, however, such as "When

are you going to get your first truffle?" (from a colleague), "Ideas are two a penny" (from a manager), and "We are not convinced that further funding is justified" (from a funding body), were, to say the least, unhelpful.

Gordon Brown and I worked for the New Zealand Ministry of Agriculture and Fisheries and have known each other since those early days. I wasn't to meet Alessandra Zambonelli until a 1988 conference in Spoleto, Italy. I began to collaborate with her group in the early 1990s, by which time she had already firmly established her place in truffle research in Italy. We hope you enjoy reading our book as much as it has given us pleasure in writing it.

To help make the text flow more smoothly, we have abandoned the normal scientific convention of using reference citations to fortify every point we make. For those readers who would like to delve deeper into the literature, however, we have provided a comprehensive reference list at the end of the book. In addition, we have listed the specific reference citations, paragraph by paragraph, on Truffles & Mushroom's Web site (www.trufflesandmushrooms.co.nz).

IAN R. HALL

ACKNOWLEDGMENTS

W E ARE INDEBTED TO OUR colleagues in France, Italy, Spain, Australia, the United States, and New Zealand who provided information for this book and to the pioneers of truffle cultivation in France, Italy, and Spain. We are extremely grateful to the trufficulteurs and tartufai in France, Italy, Spain, New Zealand, and Australia for their knowledge and assistance. Our thanks also go to the technical staff at Invermay Agricultural Centre and the University of Bologna, both past and present, and the staff of the public and university libraries in Dunedin, Bologna, and Chichester.

We are very grateful to the Tofani family and Foto Paci (Fossombrone) for providing us with photographs and permission to photograph inside Tofani Tartufi; Pierre-Jean Pébeyre for permission to use the illustration of canning truffles and to photograph inside Pébeyre S.A., Cahors; Dino Grammatico for his drawing of terracing steep truffières; Charles Lefevre of New World Truffières for his photographs of North American truffles; Wang Yun for his photographs of truffles from China; Christina Wedén and Eric Danell for permission to use the photograph of the first cultivated Burgundy truffle in Sweden; Juliet Fowler for allowing us to use the image of her finding truffles by sniffing the ground in Brian and Colleen Bassett's Truffière Croix du Sud near Opotiki, New Zealand; Joanna Dames for her photograph of the Kalahari truffle; Christine Fischer and Carlos Colinas for their micrograph of germinating truffle spores; Jim Gerdemann for his

micrograph of an arbuscle; Julian Mitchell and Alga Zuccaro for the illustration of the ribosomal gene cluster; Joan and John Parker for their photograph of hunting truffles with a pig; Marcello Pagliai and Fabio Primavera for permission to use the section of an Italian white truffle soil; Anna Maria Pisi for her micrograph of a *Tuber indicum* spore; Marco Morara for his photograph of Italian white truffle; Rita and Colin Roberts for providing the antique truffle bottle; Geoff Stephens for helping maintain English traditions and the face of the English countryside; Jeff Weston for his photograph of copper deficiency in his bianchetto truffière; the National Soil Resources Institute of Cranfield University, England, for permission to use part of the U.K. soil map; and Peter Harris, Patricia and Gordon Nelson, PowerFarming, Agrigarden Distributors, and Hirepool in New Zealand for permission to use photographs of their machinery.

Our thanks also go to Thomas Burke, Charles Montat, and Jim Trappe for helpful discussions; Michel Courvoisier for data on French truffle production; Lahsen Khabar for information on desert truffles; Chris Macquet for information on the Kalahari truffle; and the Miss E. L. Hellaby Trust for some timely financial assistance at the start of research on the cultivation of edible ectomycorrhizal in New Zealand.

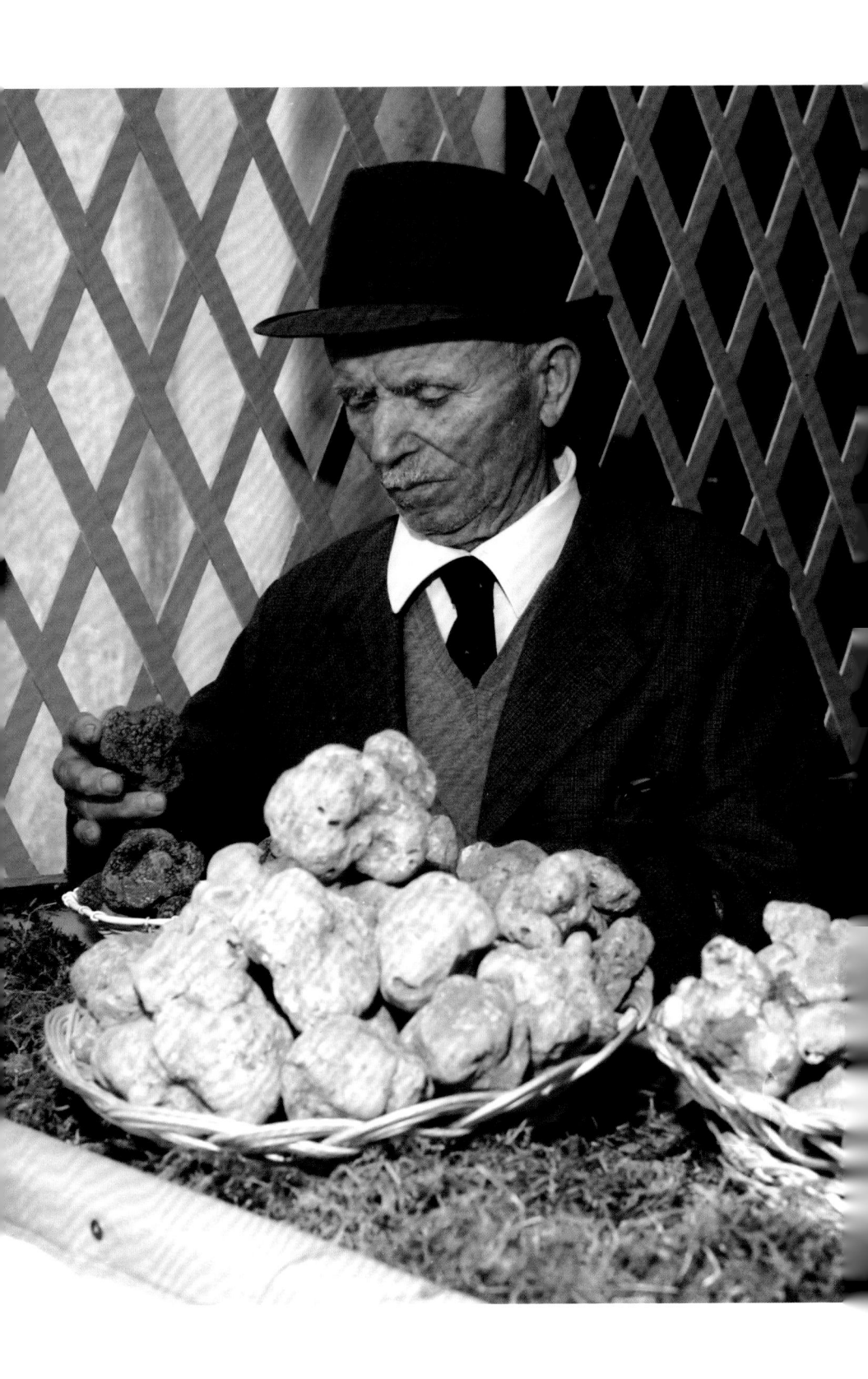

ONE

From the Past Comes the Present

T HE EARLY HISTORY OF TRUFFLES can only be garnered from the often all too brief references ascribed to luminaries of the past, and it is likely these recorded comments referred to just a few of the many edible truffles now known to exist throughout the world. Of these, desert truffles (*Eremiomyces*, *Kalaharituber*, *Terfezia*, and *Tirmania*) are still prized in the Middle East, North Africa, and by the Bushmen of the Kalahari. However, it is the Périgord black truffle (*Tuber melanosporum*) of France and the Italian white truffle (*Tuber magnatum*) that dominate today's truffle world.

The harvesting and marketing of truffles is a world that retains some of the mystery and intrigue of the past, a world that could easily be mistaken for the realm of fiction, with its record of rivalry, skulduggery, and parochialism. Whether the world's best truffles are found in Italy's Piedmont or France's Périgord is fertile ground for debate, particularly between the Italians and French. But what cannot be debated is the global demand for both, a demand that cannot be met, one that pushes prices through the ceiling, encourages trade deception, and for nearly two centuries has fuelled efforts at cultivation.

Alfred Tofani Foto Paci

15

TOP The Italian white truffle (*Tuber magnatum*) is one of the world's most expensive foods. M. Morara

ABOVE Because the Périgord black truffle (*Tuber melanosporum*) is harvested at a different time of year than the Italian white truffle, the two do not compete with each other in the marketplace. I. R. Hall

The mystery of truffles that intrigued some of the great thinkers of the past is today occupying the research efforts of mycologists around the world. While fanciful theories have been replaced with sound scientific knowledge, the holy grail of truffle research—their artificial cultivation— is still problematic. Such is the interest in truffles, as well as their gastronomic and trade possibilities, that the focus of research has spread from the Périgord black truffle and the Italian white truffle to include a range of subterranean relations, several of which are well regarded in culinary circles.

The Burgundy truffle (*Tuber aestivum*) has excellent gastronomic qualities, which has led to its successful cultivation in France. Being less fussy as to soil, host tree, and climate than its illustrious cousins, the Burgundy truffle is probably the most common edible truffle species in Europe. *Tuber aestivum* occurs naturally from the Mediterranean Basin in the south to the island of Gotland off the east coast of Sweden in the north and from the Atlantic coast in the west to an eastern European limit as yet undetermined. The Burgundy truffle has also been found in packs of truffles exported to Europe from China.

Several species of truffle, including the most important *Tuber indicum*, are traditional foods and are used as tonics by the Yi and Han people in China. Local names such as *wu-niang teng* (no mother plant) reflect the confusion of early European thinkers as to how truffles were formed. Generally, these and other Asiatic truffles have not received a good reception in the trade, despite exports to Europe increasing dramatically since the early 1990s.

Other truffles, such as Italy's bianchetto (*Tuber borchii*), have important local markets. While only truly valued in Italy between Ferrara and Ravenna, bianchetto (whitish truffle), so called to distinguish it from the more expensive *Tuber magnatum*, has excellent culinary credentials and is gaining in gastronomic appreciation. In the United States and Canada, the garlic-odoured Oregon white truffle (*Tuber gibbosum*) is abundant in the Douglas fir forests that extend from San Francisco northward to British Columbia. This species is thought by some U.S. enthusiasts to be the equal of the European truffles. Such claims notwithstanding, the Périgord black truffle and the Italian white truffle remain pre-eminent.

Writing about truffles in 1693, Sir Tancred Robinson was probably referring to the Burgundy truffle (*Tuber aestivum*). The one shown here is the first cultivated Burgundy truffle to be harvested in Sweden. C. Wedén and E. Danell

Bianchetto truffles (*Tuber borchii*) are excellent to eat, provided they are not mixed with inferior species such as *Tuber maculatum*. I. R. Hall

Regions of Renown

The Italian white truffle is harvested primarily in Italy from Piedmont in the northwest and Emilia–Romagna in the northeast to Calabria in the south. Although the neighbouring provinces of Asti and Alessandria are equally esteemed sources and the Marche region along the Apennine chain in central Italy is more productive, Alba is the mecca for *Tuber magnatum*, named by the 18th-century Piedmontese mycologist Vittorio Pico. Occupying the site of the Roman Alba Pompeia on the banks of the Tanaro River 60 km southeast of Turin, the ancient market town of Alba is at the centre of one of Europe's premier truffle territories. With a pungent and pleasant scent and peculiar but superb flavour, the white truffle of Alba, known locally as *tartufo bianco d'Alba* or *tartufo bianco pregiato*, is reputed to be the most splendid of all. The Italian composer Gioacchino Rossini referred to this species as "the Mozart of mushrooms."

Similar epithets have been lavished on the black truffle of Périgord (*Tuber melanosporum*), or *truffes du Périgord*, named for the old French province situated east of Bordeaux on the southwestern rim of the Massif Central. Although the Drôme and the Vaucluse in France, Umbria in Italy, and parts of Spain may claim to be larger producers of *T. melanosporum*, it is Périgueux, centre of the Périgord region and capital of the Dordogne, which is regarded as the black truffle capital of the world.

Piedmont and Périgord are similarly a juxtaposition of land sculpted by centuries of cultivation and untamed ruggedness—secluded wooded valleys and ravines or stark plateaus. Both carry reminders of the passage of history, of a great and failed empire. In these centres, truffles are the foundation of a culture steeped in folklore, tradition, mystery, and secrecy. It is a culture expressed in a simple but celestial cuisine and in festivals and fraternities, such as the Ordine dei Cavalieri del Tartufo e dei Vini di Alba and the Confrérie de la Truffe Noir et de la Gastronomie.

The landscape of Piedmont ranges from severe alpine valleys to flatlands, with a buffer of undulating hills between. The marriage of terrain and climate favouring truffle production is found in the Langhe and Roero Hills, respectively to the south and north of the Tanaro River in the upper

Po River Basin, and the Monferrato Hills further east, in Asti and Alessandria Provinces. In this rolling to steep terrain, wooded with oak, hazelnut, and linden growing in light calcareous soils, the *trifolau* (truffle hunter) with his dog mines the white fungal diamonds of Piedmont in the first dim light of cold, foggy mornings in late autumn and early winter. These phantoms of the night may later gather on the Via Maestra and surrounding piazzas in Alba, where, unusually, truffles are on open display, the air redolent with their aroma. Characteristically, truffle dealing is kept clandestine by a tradition of reticence, secrecy, and trickery and a desire to avoid the *fisco* (taxman).

Around Alba, the capital of the Langhe region, the woods and pastures of the upper reaches of the range give way to an idyll of hills and valleys almost entirely dedicated to the cultivation of wine grapes and hazelnuts. To the north, the Roero Hills are more compact, more demanding, yet the patchwork of vineyards, grain-fields, and woods are proof of succulence in a difficult and rugged land, furrowed by ravines. Alba's narrow streets are lined with elegant shops. The city's San Lorenzo Cathedral dates from 1486, and the civic museum is noted for its collections of Roman and prehistoric relics. Only three sentinel towers—in Via Cavour, Piazza Duomo, and Via Vittorio Emanuele II—remain from the 100 incorporated in the defensive walls built in the Middle Ages. In the surrounding countryside, hilltop citadels, fortified towns, and medieval castles are testament to sieges of times gone by. The magic views of vineyards are reminders that, in addition to white truffles, the region produces some of Italy's best wines.

In their turn, the limestone plateaus of Périgord and the neighbouring former province of Quercy to the southeast are ideally suited to the production of Périgord black truffles. Cut by deep, fertile river valleys, the country is well known for its caves and natural limestone shelters. Périgord may be the oldest settled region in France. Its two main rivers, the Vézère and Dordogne, are lined by spectacular bluffs that once sheltered Cro-Magnon cave dwellers. These Stone Age people found the climate suitable and food, in the form of bison and deer, plentiful. Following the example of wild animals, they may have dug for truffles to be baked and eaten in painted caves, such as those found in the Dordogne valleys of Lascaux and Les Eyzies.

In Périgueux the past blends pleasantly with the present. In Puy-Saint-Front and Cité, the old centres of the town, the relics of prehistoric civilization and the Roman occupation are evident. The Byzantine cathedral dates from 984 A.D., and the streets are bordered by Gothic chapels and Renaissance houses. Sarlat-la-Canéda, 65 km to the southeast, is a beautifully preserved mediaeval town. The nearby fortified hilltop village of Domme is testament to the troubled times that followed the fall of the Roman Empire. In contrast to the timeless and picturesque villages, the prosperous towns of Périgord bustle with activity. Stone churches and chalk-white farm houses grace the countryside, and magnificent chateaus above the Dordogne and Vézère are reminders of the protracted Anglo-Franco struggle known as the Hundred Years War.

This southwestern corner of France is a region of contrasts. The limestone plateaus of Quercy stretch out in stark loneliness compared to the wooded plateaus of Périgord, with their white outcrops of limestone. Coppices of oaks, chestnuts, maritime pines, and juniper shrubs grow naturally in the gritty open-textured soil. In the more fertile river valleys and low-lying areas unsuited to truffle production, a variety of crops, including fruit, grapes, cereals, tobacco, and walnuts, are grown, and dairy cattle graze rich pastures. Of all these crops, however, it is the truffle, in close association with foie gras, which distinguishes the region. The Périgord black truffle grows naturally in open woodlands characterized by the absence of plant life under the trees. These bare areas, signifying the presence of the truffle fungus, are referred to as *terre brûlé* (burned ground) in France and *bruciata*, *pianello* (a flat zone), or *cava* (digging) in Italy.

Since the 1800s, there has been a steady increase in the more orderly plantings of oak and hazelnut plantations, known as *truffières* in France and *tartufaie* in Italy. Conspicuous signs warn truffle poachers these areas are protected. While the signs point primarily to the need for laws to protect a lucrative industry, their presence on the borders of cultivated plantations also indicates the progress that has been made in understanding, controlling, and exploiting the processes of nature. Attempts at taming the truffle, of ordering its growth and harvest, now span the globe.

La Grande Mystique

There has been some success in unlocking the secrets of what the French aptly refer to as *la grande mystique*, referring to past confusions over where truffles came from, the unpredictability of the harvest, and the like. This has fascinated writers, philosophers, researchers, and bon vivants for thousands of years.

Although an anonymous reference describing truffles as "mysterious products of the earth" has been reported as early as 1600 B.C., the first attributable statement was made by the Greek philosopher Theophrastus (c. 370–286 B.C.), a pupil of Aristotle. Theophrastus' observations of plant life in general were remarkable for their precocity. He established scientific terminology where none had previously existed, united many species into generic groupings, and made careful observations of seed germination and development. Writing in his *Historia plantarum*, Theophrastus described truffles as plants without root, stem, branch, bud, leaf, flower, nor fruit; neither bark, pith, fibres, nor veins. As to their reproduction, he noted what some thinkers of the time believed, anticipating the thoughts of 17th-century researchers: "There are people who believe that they are or can be raised from seed. At all events they say that [truffles] never appeared on the shore of the Mitylenaeans, until after a heavy shower some seed was brought in from Tiarae; and that is the place where they are in the greatest numbers."

Whereas the Greek physician and poet Nicander (c. 185 B.C.) wrote of "the evil ferment of the earth that men generally call by the name of fungus," the Roman orator, philosopher, and statesman Cicero (106–43 B.C.) took a more benign view in suggesting that truffles were children of the earth. Dioscorides (40–90 A.D.), the Greek physician and pharmacologist whose work *De materia medica* was the foremost classical source of modern botanical terminology and the leading pharmacological text for 1600 years, thought that the truffle was a tuberous root.

The Greeks and Romans ate and enjoyed a variety of fungi. They identified about 20 species, the most prized among the Romans being Caesar's mushroom (*Amanita caesarea*), fungi suilli (*Boletus edulus*), and tubera, desert truffles imported from Tunisia and Libya (see Chapter 4). However,

at times their identification of edible species was found wanting and poisonings were not uncommon—a problem still with us today. The emperor Claudius, described as "greedy of such meats," was thought to have been poisoned by mushrooms. His stepson and successor, Nero, famed for his cruelty and debauchery and referred to by one observer as "that angler in the lake of darkness," called mushrooms the "food of the gods" in what is thought to have been a sarcastic reference to Claudius' deification. But of all the fungi known in that period, it was the truffles to which early writers applied their attention and recorded their ideas.

Pliny the Elder (23–79 A.D.), known for his *Historia naturalis*, which was regarded as a scientific source book during the Middle Ages, marvelled at the growth of truffles without roots: "Among the most wonderful of all things is the fact that anything can spring up and live without a root. These are called truffles (tubera); they are surrounded on all sides by earth, and are supported by no fibres or hairlike root-threads (capellamentis); nor does the place in which they are produced swell out into any protuberance or present any fissure, they do not adhere to the earth." Pliny described the truffles' general habitat as "dry sandy places which are overgrown by shrubs." As to their composition, he wrote, "they are surrounded by a bark, so that one cannot say they are altogether composed of earth, but they are of a kind of earthy concretion; in size they are often as large as quinces and weigh as much as a pound. There are two kinds: one is sandy and injures the teeth, the other without any foreign matter (sincera) they are distinguished by their colours being red, or black, or white within. . . . In their being liable to become rotten, these things resemble wood." In the end, however, Pliny was somewhat confounded: "Now whether this imperfection of the earth (vitium terrae)—for it cannot be said to be anything else—grows, or whether it has at once assumed its full globular size, whether it lives or not, are questions which I think cannot easily be explained." Of their natural history and reproduction Pliny noted, "Peculiar beliefs are held for they say that [truffles] are produced during autumn rains, and thunderstorms especially which are the main reason of their growing, and that they do not last more than a year, and are at best for food in the spring. Some think they are produced from seed."

Thunderstorms also were emphasized by other writers of the time. The Roman satirist and poet Juvenal (60–140 A.D.), who stated that a shortage of grain was preferable to a shortage of truffles, linked their origin with thunder, and the Greek biographer Plutarch (46–120 A.D.) thought them a conglomeration produced by the action of lightning, warmth, and water on the soil. Athenaeus (c. 200 A.D.), the author of the 15-volume work *Deipnosophistoi* (The Gastronomes), thought the number and size of truffles was influenced by the number and force of thunderclaps, a view still held by the Bedouin.

The oldest surviving recipes for cooking truffles can be found in the curious and fascinating book *Apicius de re coquinaria*. This record of cooking and dining in Imperial Rome, the oldest known European cookery book, contains six recipes for cooking truffles, a recipe for a wine sauce to serve with truffles, and a method of storing truffles in sawdust. There is some doubt as to who Apicius was and whether the book was compiled by one man or is a collection put together over a period of 200 to 300 years. Many researchers believe he was M. Gabius Apicius (d. 40 A.D.), a celebrated bon vivant who lived during the reigns of Augustus and Tiberius.

There is no doubting the fascination truffles held for Roman and Greek writers and gastronomes, but the fungus with which they were familiar was not the "black diamond" of Périgord, but the desert truffle (*Terfezia*). Indeed, Pliny the Elder noted, "those of Africa are most esteemed." The Romans and Greeks would have known the desert truffles of North Africa, Turkey, the Middle East, and semidesert regions of the Mediterranean Basin. Historically, markets for these truffles were found in Damascus, Baghdad, Izmir, Aleppo, Baku, Tbitisi, and Jerusalem.

Much of what was written about truffles by the Greeks and Romans can be dismissed as fanciful. For instance, as late as 234 A.D. the Greek philosopher Porphyrius was referring to them as "children of the gods." In addition to the theory that truffles might be propagated by seed, however, it was noted that truffles were always found in the ground beneath a particular plant the Greeks called *hydnophyllon*, a species of rock rose. Had the decline and fall of the Roman Empire not ended an era of learning and intellectual curiosity in the West, this hint of a relationship between plant

and fungus may have provided the foundations for more useful observations of the truffle.

The years between the 5th and 12th century are sometimes referred to as the Dark Ages, because of the apparent extinction of culture and learning. As one writer aptly put it "night fell on the domain of the sciences." Classical Greek and Roman manuscripts existed only in ecclesiastical libraries and were largely ignored. The last recorded mention of the truffle before the long medieval silence came from S. Ambrogio, archbishop of Milan, in a letter very much appreciating a gift of large truffles made from S. Felice. Little or nothing was added to the knowledge of truffles during the Dark Ages. They acquired a sinister reputation and became known as the devil's handiwork, grown from the spit of witches.

Of that period, the French mycologist M. G. Malençon said the truffle was forgotten for more than a 1000 years and it was only towards the 14th century, when the time of barbarism was at last over, that the truffle returned to its place on the tables of the wealthy in Italy and France. Testimony to that return can be found in a reference to truffles by the Italian poet Petrarch (1304–1374) in his ninth sonnet of the Rime: "Not just the world apparent from outside / the river-banks and hills, receives the sun, / but there within where day is never born / earth's moisture is made pregnant with the light, / the very reason for such fruits as these" (Petrarca 2000). Looking back on the lack of scientific activity in the Middle Ages, the French botanist Gaspard Adolph Chatin wrote in 1892, "The character of that time is in the total abandonment of speculative views about the origins and constitution of the truffle which does not occupy a gastronomic point of view."

Such speculation returned in the 16th century, no longer confined to Greece and Italy but much influenced by the classical writers. In 1552 the German herbalist Jerome Bock wrote, "Fungi and Truffles are neither herbs, nor roots, nor flowers, nor seeds, but merely the superfluous moisture of earth, of trees, of rotten wood, and of other rotting things. This is plain from the fact that all fungi and truffles, especially those that are used for eating, grow most commonly in thundery and wet weather." Some light was cast on what truffles were and how they should be classified by Alfonso

Ciccarello, a doctor from Perugia, when he hypothesized in his *Opusculum de tuberibus* (1564) that truffles, like mushrooms, were the fruits of a fungus.

However, fancy and confusion still abounded. In 1573 the famous Italian naturalist Mattioli still described truffles as "roots without stalks." Similarly, in 1597 the English herbalist John Gerard was calling the truffle a "tuberous excrescence" with "neither stalks, leaves, fibres nor strings annexed or fastned unto them." In 1623 the Swiss botanist Gaspard Bauhin considered fungi to be "nothing but the superfluous humidity of soil, trees, rotten wood and other decaying substances." Indeed, when an edible white tuber from South America—the potato—was introduced to Europe via Spain at first it was thought to be a variety of truffle. The exotic tuber was called *tartufo* in Italy and *truffe* or *truffe rouge* in France until its true relationship was established.

A major breakthrough toward understanding the true nature of truffles came with the first observation of spores by the Italian natural philosopher Giambattista della Porta in 1588. In his *Phytognomonica* he wrote, "From fungi I have succeeded in collecting seed, very small and black, lying hidden in oblong chambers or furrows extending from the stalk to the circumference, and chiefly from those which grow on stones, where, when falling, the seed is sown and sprouts with perennial fertility. . . . in truffles, a black seed lies hidden. On this account, they come forth in woods where they have frequently been produced and have rotted away." Despite ongoing misconceptions and confusion, della Porta's observation did not pass unnoticed. In 1640 the English botanist John Parkinson quoted him as saying, "Under the outer skinne, certain small blacke seed . . . whereby it not only propagateth it selfe, where it is naturall, but as it hath been often observed, there hath some of them growne where the parings of them have beene cast" (quoted in Ramsbottom 1953).

The British scientist Robert Hooke, writing in 1665, ruled out the presence of seeds, and in 1693 the English naturalist Sir Tancred Robinson, referring to truffles as a "delicious and luxurious piece of dainty," admitted he was in the dark: "What these trubs are, neither the Ancients nor Moderns have clearly informed us, some will have them Callosities, or

Warts bred in the Earth: Others call them subterraneous Mushrooms." Robinson's truffles were probably *Tuber aestivum*, the Burgundy or summer truffle.

Professor Richard Bradley, the head of the botany department at Cambridge University, wrote in the appendix to his *Dictionarium botanicum* (1728) that he had "taken some Pains to search for them in England, and have found them in many Places, in Woods especially, in Surrey, Middlesex, Kent, Essex, in Hertfordshire, and Northamptonshire, and I guess, we have very few old Woods in England without them." He went on to note, "The species of truffle found in England is not the same as either of the two principal edible species found in France and Italy, although still worth eating." Interestingly, Bradley provided one of the earliest records of an intention to cultivate truffles, "They are very plenty in our Woods in England, as I understand by several who have found them this Summer by my Directions, and I believe will be much more so, since several Gentlemen have followed my Advice in propagating them."

Towards the end of the 17th century, the idea that truffles were propagated by seed was again advanced, this time by the English naturalist John Ray, known for his work on the classification of fungi into terrestrial, arboreal, and hypogeous categories. That della Porta's observations and Ray's theories of propagation had gained some currency was evident in the publication of a popular gardening manual, *The Practical Kitchen Gardener*, by Stephen Switzer in London in 1727. Switzer wrote, "It is a pity that we can't as yet find out a method of propagating these for much desired dishes; perhaps there might be a method of doing it by the procuring of the earth where they grow, which certainly contains some seminalia [seeds] or fragments of those tuberous roots which when transplanted out might grow with us . . . Mr. Ray says of them, that the roots are of an unequal globular figure; that they grow in sandy ground, and under trees. . . . the place of their growing is discovered by certain chasms or clefts, that are discovered in the superficies of the earth."

Further observations that helped, in time, refute the widely held view that truffles and other fungi originated by spontaneous generation were made in the first decade of the 18th century by French botanist and physi-

cian Joseph Pierre de Tournefort and the pharmacist and botanist Claude-Joseph Geoffroy. De Tournefort was a pioneer in systematic botany whose system of plant classification represented a major scientific advance. He observed and described the spores of truffles and is reputed to be the first person to describe how they might be successfully cultivated. Geoffroy was a follower of de Tournefort, and it was Geoffroy who helped settle the botanical confusion surrounding the truffle when he classified it among mushrooms in a 1711 paper entitled "Vegetation de la Truffe," thus confirming Ciccarello's hypothesis made almost 150 years previously.

These observations were confirmed by Pier Antonio Micheli, the Italian botanist who provided the first definitive description of "seeds" (spores) in truffles. He noted that small clusters of "seeds" (spores, or more correctly "ascospores") developed inside membranous transparent sacs (asci), and the 1729 publication of his *Nova plantarum genera* is regarded as the birth of mycology.

Micheli's view that truffles, and fungi in general, were propagated from spores was debated for the next century but drew increasing support. In 1788, the English naturalist James Bolton wrote, "The plants which now compose the Order Fungi, were formerly supposed to be of equivocal generation, the sport of nature, the effect of Putrefaction, or the brood of Chance; but that they owe their original to the seeds of a parent plant,

In 1729 Pier Antonio Micheli saw truffle spores (red arrow) forming inside transparent sacs (asci; blue arrow) within truffles. These Périgord black truffle (*Tuber melanosporum*) spores are ornamented with spines that are slightly curved at the tips, a characteristic of this species. I. R. Hall

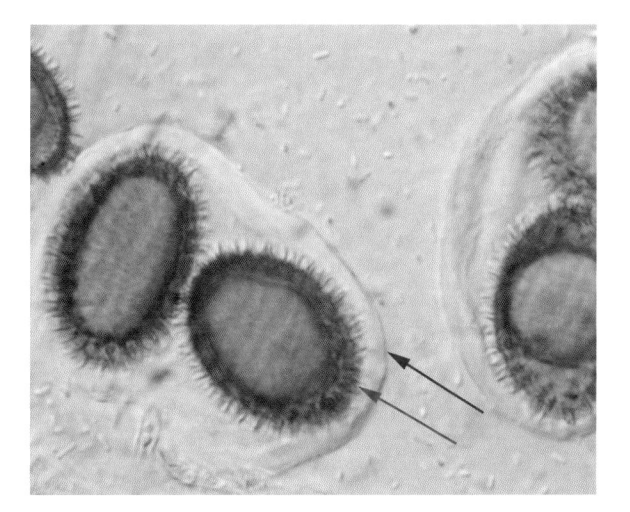

From the Past Comes the Present

is now well known." In 1791, Pierre Bulliard saw truffle spores clearly but failed to note Micheli's transparent sacs. Bulliard concluded that the spores were connected to the surrounding tissue by delicate, threadlike "umbilical chords," or fibrils, until they reached maturity: "If at this time [maturity] the seeds are observed through strong lenses they can be seen to perfectly resemble the truffle in which they developed; like it, they have a rounded or slightly elongated shape and their surface, standing up in points, appears to have been chiseled; they are also, like it, a brownish black; finally they differ only absolutely in their size . . . the result is that they cannot be regarded as seeds but as completely formed little truffles" (quoted in Malençon 1938). Bulliard mistakenly concluded that truffles were viviparous plants (that is, offspring that germinate while still attached to the parent plant). This view, due to the regard in which he was held, was widely accepted.

Bulliard's regression on the work of Micheli over half a century before was influential until the published works of Carlo Vittadini (1837) and the Tulasne brothers (Tulasne 1851) put the scientific study of truffles on solid ground. Such was the extent and excellence of the work performed on fungi in general by Edmond Louis-Rene Tulasne and his younger brother Charles, that they are considered the founders of modern mycology.

But such scientific developments struggled to find acceptance among a wider population bewitched by the romance of fairy tales, songs, and poems. One legend told in the Périgord region is about an old woman, tired and hungry, losing her way in the woods. At last she stumbles upon a tumbledown house, the home of a man equally poor and old. He invites her in and offers her his meagre meal of charred potatoes, cooked in the coals of a dying fire. The old woman is deeply touched by his generosity and sits down to peel the potatoes. Suddenly she is transformed into a beautiful fairy. "Lo, do not be alarmed old man," she says. "I am the fairy of the woods. You are a kind and noble person and from these poor potatoes, which you have humbly shared with me, will come the end of your trials and tribulations." Before his eyes the charred potatoes are transformed into richly flavoured truffles. Despite becoming wealthy and respected throughout the region, the old man continues to be kind and helpful to all those less fortunate than himself, but the same cannot be said for his children. They grew

up spoiled and lazy, and when many years later the good fairy returns, disguised as an old woman, they refused her hospitality and food. To punish them, the fairy buries all of the truffles underground and turns the selfish children into pigs to root them out.

Theories not far removed from the realm of fairy tales and little different from those of the Greeks and Romans were still being advanced in the mid-19th century, despite the advances of science. Variously, the truffle was seen as a disease of the root system, a growth from the sting of a fly on the root of the host tree, or a result of the fermentation of the soil. In 1831 Vittadini described more than a dozen truffles in the work *Monographia tuberaceum*, but it was the French botanist Gaspard A. Chatin's book of 1868 (revised in 1892) that contained a wealth of factual information and refuted the many imaginative but erroneous theories that had been put forward. Chatin cautiously indicated that the truffle might be cultivated. He believed truffles were reproduced from spores and formed a symbiotic relationship with the host tree, and that the spore, rather than the tree, determined the truffle's species. All of these assumptions have since been proven true.

The Golden Age of Truffles

The unravelling of some of the biological mystery of truffles had been paralleled by a dramatic growth in their culinary stature, from being a food source for peasants during the Dark Ages to becoming an obsession in higher circles of society, what French savant of haute cuisine Jean Anthelme Brillat-Savarin (d. 1826) described as "the jewel of cookery."

Brillat-Savarin's time marked the beginning of what might be dubbed the "golden age of truffles," particularly in France, where a heightened appreciation for them was more readily matched by supply than at any other time in their history. This ready availability was indicated in recipes such as that requiring a "dozen fine black truffles" in Mrs. Beeton's first cookbook (1861) and by the response of food writer Maurice Edmond Sailland, better known by his pen name Curnonsky and dubbed the Prince of Gastronomy, who, when a Parisian hostess asked how he liked his truffles, replied, "In great quantity, madam, in great quantity."

Truffle appreciation among the nobility of France appears to have grown, at least in part, as a result of Italian influences, the earliest of these being the move of the papacy from Rome to Avignon between 1309 and 1377. Towards the end of this period, Jean de Valois, the first Duke of Berry and third son of King Jean II of France, is said to have developed a taste for truffles. He featured them in banquets hosted by himself and his wife, Jeanne d'Armagnac, and introduced Charles V to truffles in an attempt to make an impression. This endeavour may have been limited by a lack of culinary sophistication, as one reference to preparing and serving truffles at the time indicated that "after being preserved in vinegar, they were soaked in hot water, and afterwards served up in butter" (Lacroix 2004).

Progress in the culinary aspects of truffles was also attributed to Italian influence. In 1533 Catherine de Medici came to France from Florence to marry the Duke of Orleans, the future King Henry II and son of François I. She is credited with introducing the art of sophisticated cookery to France, and with it a taste for such Italian delicacies as truffles, sweetbreads, and artichokes.

While there is some suggestion the truffles referred to in these examples were either the Italian white truffle or more lowly regarded species, such as bianchetto or the summer or Burgundy truffle, there is little doubt that it was the Périgord black truffle (*T. melanosporum*) being referred to when the *Bulletin of the Archaeological Society of Périgueux* reported that truffles were served at the banquets of French nobles during the 16th-century reign of François I, their exotic nature making them a luxury product and a sign of wealth for centuries to come.

In emphasizing the long interval between the Roman era and the truffle's rediscovery during the Renaissance, Brillat-Savarin stated, "for I have read several old cooking books in which no mention is made of it." Presumably he is talking of the years prior to the 17th century, it being almost inconceivable that Brillat-Savarin would not have been aware of François Pierre de La Varenne's *Le cuisinier françois*, published in 1651 and considered to be the founding text of authentic French cuisine. La Varenne, chef to Louis XIV, featured truffles in more than 60 recipes and is credited with being the first person to pair foie gras with truffles. Despite La Varenne's

undoubted influence, an appreciation for truffles remained, paradoxically, the preserve of the privileged and the peasant. The truffle's rise to the heights of food fashion was still more than a century away, and by the mid-19th century, it was safe to say the fame of the truffle was as its zenith.

Magical Powers and Medicinal Virtues

Reinforced by the weight of religious belief and utterance, magical powers and medicinal virtues have long been attributed to the truffle. A *hadeeth*, a narration from the prophet Muhammad states, "The truffles are from Al-Mann (that is, it is given from Allah without seed or labour) and its water is a remedy for the eye," a message reiterated by the Islamic physicians Ishaq ibn 'Imran (Tunisia, c. 900) and Abu'l-'Ala' Zuhr (Andalusia, c. 1100). At around the same time the eminent Persian doctor Avicenne (Ibn Sina) was recommending truffles as a treatment for weakness, vomiting, pains, and for healing wounds. A belief in the potency of truffles appears to have been common across religions at that time. For instance, Pope Gregory IV, head of the Catholic Church from 827 to 844, is said to have eaten truffles for strength in his battles against the Saracens, who invaded Sicily during his reign.

The truffle has been revered for its ability to cure gout and used in syrup as a source of energy. According to Stephen Switzer (1727), "besides the uses of this root in cookery, I can't but observe from Cardan [Girolamo Cardano, 1501–1576], in his book 'De varietate rerum cap 28' that when it is boil'd and used plaisterwise [as a poultice], in all quinsies [tonsilitis] and soreness of the throat, that it has reliev'd those that have been at the point of death."

Since the earliest times the truffle has been considered an aphrodisiac. Aristotle referred to the truffle as "a fruit consecrated to Aphrodite," and the Roman physician Galen warned that over indulgence in truffles could lead to voluptuousness. Napoleon and the Marquis de Sade are reputed to have used them as a sexual stimulant, and Louis XV's mistress, Madame Pompadour, fearful of being unable to meet the amatory demands of the king, experimented with a diet of vanilla, truffles, and celery, intended to

"heat the blood." Georges Sand, the famous campaigner for women's right to free love, wrote "the truffle is the black magic apple of love." French author and playwright Alexandre Dumas called the truffle "the holy of holies," and writer Prunier de Longchamps warned priests and nuns that the consumption of truffles was inconsistent with a vow of chastity.

Such beliefs were not confined to France and Italy. When the English diarist John Evelyn was introduced to truffles in France in 1644, he found them "an incomparable meate," but more than 50 years later, in his Acetaria, primly referred to them as "rank and provocative Excrescences." Switzer stated that truffles were "very effective in venereal embraces." He mentioned "a kind of them . . . observ'd near Furstenwald, that resembled the testicular parts of a man," termed *scroto denudato*, and also "tubera cervina; fabled to be rais'd from the genitals of a stag, to be found at Trenzinum, a noble city of Hungary." Such remarks suggested the erotic connection was widespread in Europe.

Brillat-Savarin noted a widespread belief that "truffles are conducive to erotic pleasure." In the 1820s Brillat-Savarin took it upon himself to verify the truth or otherwise of such beliefs. After much consultation with reluctant ladies, "For all the replies I received were ironical or evasive" and men "Who by their profession are invested with special trust," concluded that "the truffle is not a true aphrodisiac but in certain circumstances it can make women more affectionate and men more attentive." This is a reasonable conclusion one might suppose, for in the right setting, with agreeable companionship, and stimulated by thoughts of the truffle's rarity, its exquisiteness and expense, there is every likelihood such a meal would be a sensual experience.

TWO

Science to the Rescue

THE POPULARITY OF TRUFFLES reached a peak in the late 19th century. Brillat-Savarin noted that merchants, aware of their popularity, stimulated the market by paying good prices at source and using the fastest transport available to send them to Paris. With a ready market came a desire to increase supply, and attention focused on the truffle harvester's holy grail—artificial and systematic cultivation.

The first step toward cultivation was made in the early 19th century with the discovery of a haphazard method of planting acorns near truffle-infected parent trees. In their book *Mushrooms and Truffles* (1987), Singer and Harris described the invention by Joseph Talon of the indirect culture of Périgord truffles: "Talon was a humble peasant of the 'Hameau des Talons,' Vaucluse. He planted acorns on a piece of stony siliceous earth, and was surprised to be able to harvest truffles under the young trees a few years afterwards. An observant and thrifty man, he proceeded to buy worthless land, disseminated there the acorns from his own plantation of 1810 and, keeping his discovery as secret as he knew how, was able to make some profit on the truffles which appeared in his successive oak plantations. When old, father Talon gave away his secret . . . to a certain truffle mer-

chant and friend of the Talons, by the name of Auguste Rousseau, . . . who sent the first lot of cultivated truffles to an exposition in the capital, Paris, in 1855, and turned propagandist for Talon's method of truffle growing."

Despite an earlier report on truffle cultivation in a letter to Giacomo Sacchetti, the secretary of the Siena Academy, that was published in 1807, the method is generally attributed to Talon. That the method was used to establish vast plantations in the late 1800s and was still being used in the second half of the 20th century is testimony to its success. Truffle production increased markedly during the second half of the 19th century, driven partly by farmers wanting to grow alternative crops in vineyards laid waste by the arrival of the *Phylloxera* aphid in the mid-1800s and diseases that killed silkworms in southern France, rendering large areas of Chinese mulberry trees useless. By 1890 there were 750 square kilometres of truffières in France, and Gaspard Adolphe Chatin (1892) gauged production to be 1500 tonnes in 1868 and 2000 tonnes by 1890. Although there is some suggestion that these figures may have been exaggerated, with actual production perhaps as low as 1000 tonnes, clearly the end of the 1800s was the golden age of truffle production.

Mycorrhiza: An Intimate Relationship

In 1885 Albert Bernhard Frank, while working on a commission from the King of Prussia to develop more reliable methods to cultivate truffles, became sidetracked and recognized an intimate relationship between the fungi and plant roots. Giuseppe Gibelli had identified the relationship two years earlier while working on chestnut ink disease, but it was Frank (1877, 1888) who coined the termed "Mykorrhizen" (from the Greek *myco*, fungus, and *rhiza*, a root; in English "mycorrhiza") to describe the symbiosis. This serendipitous breakthrough was to eventually explain how Talon's technique worked and was to be the salvation of the truffle industry in the late 20th century.

Over the decades following Frank's discovery scientists realized that, with a few notable exceptions such as the brassicas (Brassicaceae), nettles (Urticaceae), and the convolvulus family (Polygonaceae), the majority of

higher plants form mycorrhizas. While mycorrhizal fungi actually *infect* the roots of their host plants, the relationship is generally beneficial because the fungus supplies the plant with additional nutrients such as phosphorus. The fungus does this by sending out a network of fine threads called *hyphae* that penetrate areas of soil that the plant is not able to exploit. In exchange, the plant provides the fungus with nutrients, such as carbohydrates, and a place to live.

Mycorrhizal fungi have been around since the Ordovician period, nearly half a billion years. Thus, it is hardly surprising that most are now so specialized that they cannot survive unless in contact with their host plants. Many plants have also become equally dependent on mycorrhizal fungi and without them become stunted and yellow, often due to a lack of phosphorus. It is because of this close relationship that plants raised by your local nurseryman in a mycorrhiza-free, high-fertility potting mix may fail to survive after transplanting to relatively infertile situations such as your garden—but that is another story.

Types of Mycorrhizas

There are several types of mycorrhizas and, unfortunately, all have rather cumbersome names, such as arbuscular mycorrhiza, ectomycorrhiza, and ericaceous mycorrhiza. The ectomycorrhizal association has probably been around for a mere 100 million years, and thus these fungi are considerably younger than the arbuscular type, the oldest of the mycorrhizal fungi. In arbuscular and ericaceous mycorrhizas the fungus actually gets inside the cells of the outer layers of the roots, whereas ectomycorrhizal fungi wrap around the outside of the host plant's fine roots just like the fingers of a glove. Fungi that form one kind of mycorrhiza are quite different from those that form another, so generally ectomycorrhizal fungi cannot produce an arbuscular type of mycorrhiza and vice versa. Some oddities include the ectendomycorrhizas and terfezoid mycorrhizas formed by desert truffles.

Most of the trees that dominate the forests of the Northern Hemisphere, including birches (Betulaceae), oaks and beeches (Fagaceae), many

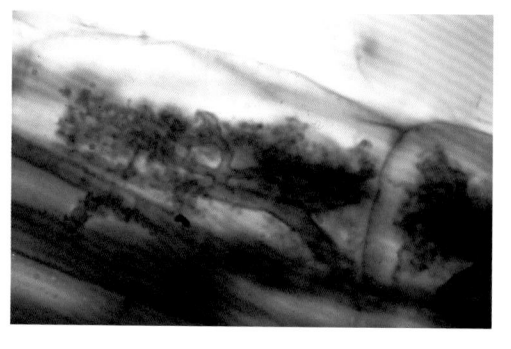
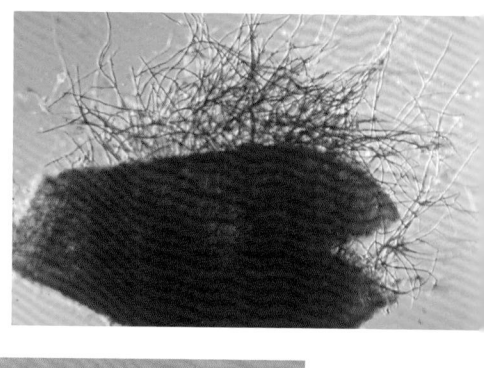

TOP LEFT An arbuscle (meaning "little tree") produced by an arbuscular mycorrhizal fungus inside a root cell J. Gerdemann

TOP RIGHT Blue stain and a special lighting called Nomarski interference has been used to show the external hyphae (cystidia) on the surface of a Périgord black truffle (*Tuber melanosporum*) mycorrhiza. Note that the tip of the mycorrhiza was split when it was flattened to take the photograph, and the root tip is obscured by the fungus. I. R. Hall

ABOVE Dense clusters of Périgord black truffle (*Tuber melanosporum*) ectomycorrhizas of common oak (*Quercus robur*) in a heavily limed, very free draining, volcanic ash soil in New Zealand. The fungus covers each root tip of the host plant just like the finger of a glove. I. R. Hall

softwoods (Pinaceae), European limes (Tiliaceae), and the tropical diptero-carps (Dipterocarpaceae), as well as the Australian eucalypts (Myrtaceae) of the Southern Hemisphere, form ectomycorrhizas. In contrast, the vast majority of cultivated crops, flowering plants, ferns, cycads, and some gymnosperms, such as the North American redwood, form arbuscular mycorrhizas with members of an obscure fungal group called the Glomeromycota. Despite the enormous ecological and economic importance of these fungi, their fruiting bodies are usually microscopic and only provide sustenance for very small animals, such as nematodes (roundworms) and beetle and fly larvae. Generally, plants within a family tend to form the same kind of mycorrhizas (see Appendices 1, 2). There are a few exceptions, however, such as the Myrtaceae family, which contains both ecto- and arbuscular mycorrhizal species. A few tree species, such as poplars and willows (Salicaceae), some eucalypts, and *Leptospermum* (Myrtaceae), have the best of both worlds as they are able to form both ecto- and arbuscular mycorrhizas.

Most mushrooms that we are likely to encounter in our local supermarket are grown on dead plant material, while a few are produced by pathogenic fungi that grow on living plants. Other delicacies, such as the black trumpet (*Craterellus cornucopioides*), the golden chanterelle (*Cantharellus cibarius*), matsutake (*Tricholoma matsutake*), and porcini (*Boletus edulis* sensu lato), like the truffles, are ectomycorrhizal mushrooms. In fact of the estimated 5000 to 6000 species of ectomycorrhizal fungi, more than 900 produce mushrooms that are considered safe to eat. Sadly, fewer than a dozen have been cultivated, although some of the techniques we describe for truffles in this chapter and in Chapters 5 and 6 have the potential of unlocking the secrets for the cultivation of other edible ectomycorrhizal mushrooms—a legacy of applied science that stems back more than two centuries.

Structure of Ectomycorrhizas

Most ectomycorrhizas have a layer of fungal tissue on the surface of the fine roots called the mantle. From the mantle, tongues of tissue run in between

the outer layers of the root to produce a three-dimensional structure called the Hartig net. This can be visualized by imagining that the outer layers of cells of the root are like the bricks in a chimney and the fungus is the mortar between them. On the outside of the mantle, fungal hyphae run out into the soil.

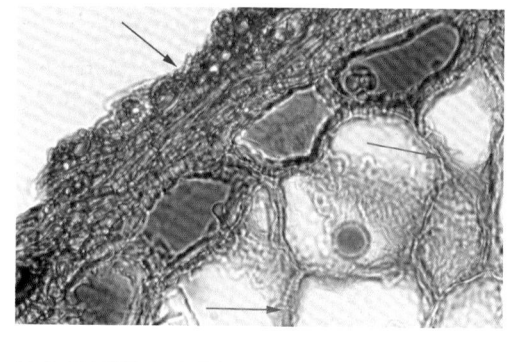

TOP A cross-section of a Périgord black truffle (*Tuber melanosporum*) ectomycorrhiza stained with blue dye. From the inside of the mantle, the thick fungal layer that covers the outside of the root, tongues of fungal material penetrate between the outer cells of the root to form the Hartig net
A. Zambonelli

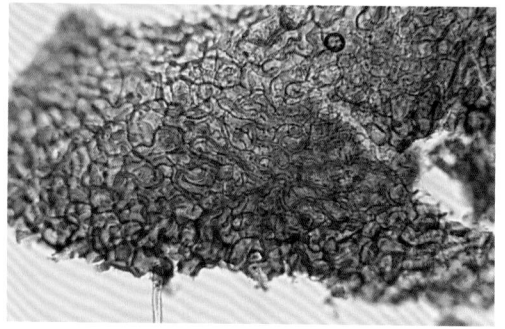

CENTER A view of the surface of a Périgord black truffle (*Tuber melanosporum*) mycorrhiza showing the intricate paving arrangement of the outer cells of the mantle. The appearance varies across species and can be used as a means of identification. I. R. Hall

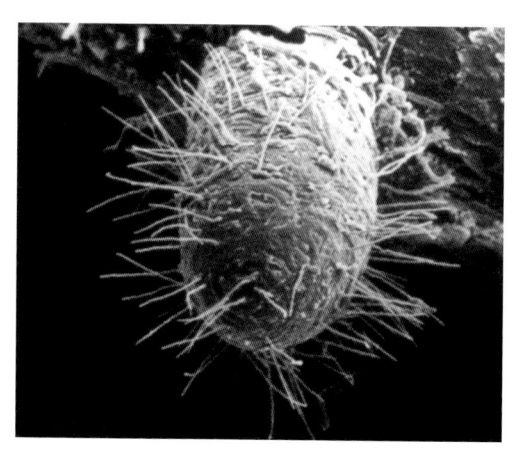

BOTTOM Electron micrograph of a bianchetto truffle (*Tuber borchii*) mycorrhiza showing the characteristic bottlebrush appearance of truffle mycorrhizas. The projecting hyphae (cystidia) of bianchetto mycorrhizas are particularly long and thin. A. Zambonelli

Saffron milk cap (*Lactarius deliciosus*) mycorrhizas are bright orange and turn green with age, which makes them fairly easy to identify. Such examples are rare, however, and the identification of most mycorrhizas requires considerable experience and the aid of a powerful microscope. Nonspecialists can easily confuse the small sausage-shaped mycorrhizas of the Périgord black truffle with similar ones formed by other ectomycorrhizal fungi, such as *Sphaerosporella brunnea*. Young, actively growing truffle mycorrhizas are covered with microscopic needlelike projections (cystidia) that radiate out from the surface, which can give them the appearance of miniature bottlebrushes. The size and shape of these, along with the structure and surface characteristics of the mantle, can be used to distinguish the various species of truffle (see Chapter 3).

Underground Warfare

In the forests of the Northern Hemisphere that are dominated by ectomycorrhizal trees, it is not unusual to find many different ectomycorrhizal fungi on the roots of a single host plant. But these fungi do not live together in peace and harmony—there is in fact intense competition among them for space on the roots and for nutrients. Gerard Chevalier put it succinctly when he described the situation as "underground warfare." Some species have evolved rapid strike tactics—infect quickly and before others can get a foothold—while others use chemical weapons to secure their share of the available space. Some, including the Burgundy and Périgord black truffle, have also evolved a tolerance to soil conditions that other species cannot survive in.

Like most wars, it is sometimes difficult to fathom which are the good guys and which are the bad ones, and the winners change depending on the conditions. Some ectomycorrhizal fungi are true symbionts. The fungi obtain a substantial proportion of their nutrients from their host, and the host plant receives a more efficient structure for absorbing nutrients from the soil and protection by having the fungus enveloping its fine roots. These benefits for the host plant far exceed the nutrient costs. However, some ectomycorrhizal fungi also have enzymes that enable them to derive some

of their nutrients directly from organic matter in the soil. Growth depressions in host plants in response to inoculation with some ectomycorrhizal fungi have been detected, and under some circumstances and at certain times the Japanese delicacy matsutake (*Tricholoma matsutake*) may even harm its host plant.

Some ectomycorrhizal fungi are not restricted to just one host. For instance, the red and white fly agaric (*Amanita muscaria*) is found on eucalypts, pines, oaks, and birches. Similarly, some host plants can form ectomycorrhizas with a range of fungi; numerous species can form a succession on the roots of a host, somewhat similar to the succession of plants one would see growing on a neglected bare patch of ground. But other ectomycorrhizal fungi have evolved a strong preference for a particular host and presumably derive a commensurate benefit from doing so. For example, in the United Kingdom about 40 per cent of ectomycorrhizal fungi are associated with just a single genus of host plants; of the large number of species associated with Douglas fir (*Pseudotsuga menziesii*), about 10 per cent are unlikely to be found with any other host. Similarly, each truffle tends to have relatively few hosts (Appendix 3). Fungi that form mycorrhizas with, for example, pines, poplars, and willows rarely form mycorrhizas with eucalypts, and eucalypt-compatible ectomycorrhizal fungi rarely form mycorrhizas with fungi originating outside of Australia. As a consequence, in Tasmania very few native ectomycorrhizal fungi have been found contaminating truffle-infected oaks, and only a handful of accidentally introduced ectomycorrhizal fungi are found associated with the Monterey pine (*Pinus radiata*) plantations of Australia and New Zealand. It was this depth of understanding of mycorrhizas, developed in the post-Frank era, that was eventually needed to reverse the collapse in the truffle industry that began at the start of the 20th century.

The Collapse of Truffle Production

With two major wars in Europe, economic devastation, and the gradual migration of people to cities during the first half of the 20th century, truffle plantations were abandoned and truffle production declined dramatically.

Present-day production palls by comparison with the harvests in the late 19th century, and several reasons have been advanced to explain this collapse. At the beginning of the 20th century, truffle growing and harvesting was cloaked in mystery, much as it has always been. The location of known truffle beds was a closely guarded secret known only to a select few. Truffles and their harvest were the preserve of men. Women, considered to be "impure," were kept away from truffle beds for fear that their very presence would strike the beds sterile. Only on his deathbed would the truffle grower pass on to his sons the secrets of truffle cultivation, or even the places where they were to be found in the wild. As many truffle growers died in the trenches during the First World War, their secrets often died with them. Their families had great trouble even finding the truffle beds, let alone knowing what to do when they did.

The typically 10-year lag between planting a truffière and harvesting the first crop, as well as the possibility that the new trees would not produce, were powerful deterrents to prospective truffle producers. Therefore, it was hardly surprising that people tended to look for more immediate and certain returns through crops that yield more quickly and predictably.

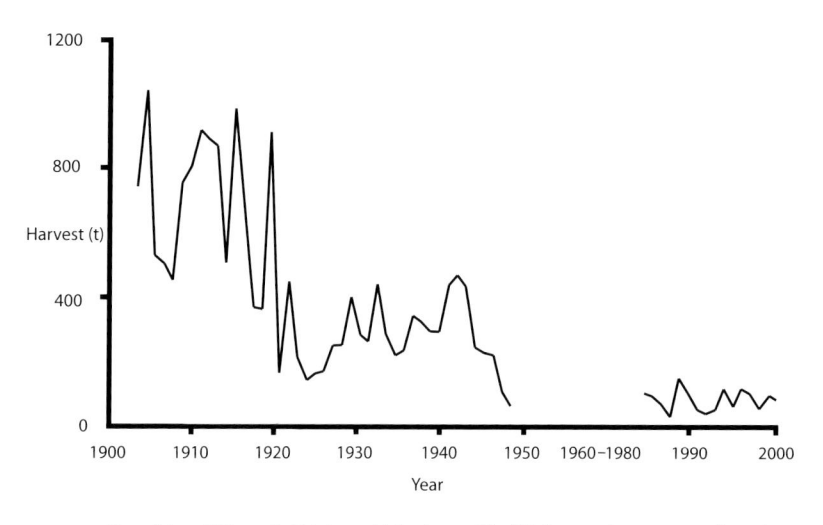

Combined French Périgord black truffle (*Tuber melanosporum*) and winter truffle (*Tuber brumale*) production, 1902–2001 Data courtesy of M. Courvoisier

Another factor contributing to the industry's decline was the shifting balance between rural and urban living. Truffles do not grow well in areas of dense vegetation. In the past, woodcutters and grazing animals had kept forests half clear, but with the drift to urban areas, large parts of France and Italy became wilder and overgrown.

Prices for truffles also fell. Prior to 1914 a kilogram of truffles cost 10 francs, but during the First World War prices fell to an all-time low of 3 francs per kilogram. Demand did not recover in the difficult postwar years. Oak trees were either cut down and the land developed for more profitable crops or landowners simply neglected the trees on their property, leaving many truffle-producing areas in chaos. By the end of the Second World War production had collapsed completely. Beginning in the 1960s, however, truffles were again highly valued, and new efforts to cultivate them had begun (Appendices 4, 5).

The forests of France are denser today than a century ago and less likely to be conducive to fruiting by the Périgord black truffle (*Tuber melanosporum*). I. R. Hall

New Techniques

Although Talon's technique for producing truffle-bearing trees was the mainstay of the Périgord black truffle industry for more than 150 years, it was badly flawed. The seedlings were exposed to all the organisms in the rooting zone of the donor plant, such as insect pests, nematodes, and faster-growing contaminating ectomycorrhizal fungi. By the 1960s the collapse of the European truffle industry was well underway, and new techniques were obviously needed to rejuvenate it. At that time, however, direct cultivation—employing a proven method of inoculating host plants with the truffle fungus rather than leaving it up to nature, as in Talon's technique—was still not possible. According to Singer (1961), "The waves of new developments, patents, control methods, and rationalizations that characterize the literature and practices of mushroom cultivation both in Europe and elsewhere during the 20th Century, have hardly touched the truffle growing business of southern France."

By the end of the 1960s, however, French and Italian scientists had devised methods for producing truffle-infected plants under controlled conditions in greenhouses by inoculating plants with pureed truffle, truffle spores, cultures, or sections of infected root. Eventually their perseverance was rewarded, and in December 1977 the first cultivated truffles were harvested at Aigremont, France. French and Italian scientists were finally able to release a long-held collective breath. The Spanish did not lag far behind, and the first truffières were established in the early 1970s, although truffles had been harvested from natural areas in Spain for some decades before that. Needless to say, the complete methodologies now used by companies to produce hundreds of thousands of truffle-infected plants annually remain closely guarded secrets. Although the general methods that can be used are now well documented, these publications are for a restricted readership.

Production of truffle-infected plants using spores

There is a considerable gap between producing a few infected plants in a small-scale experiment and producing tens of thousands of healthy plants

well infected with a truffle and free of contamination year after year. One can imagine, therefore, that there would have been some commercial reluctance to publish the details of the methods developed in the 1960s and 1970s. However, eventually a set of instructions for producing truffle-infected plants using spores was published by Mannozzi-Torini (1976) based on a series of experiments he conducted between 1965 and 1976. Subsequently Mannozzi-Torini's methods were improved by the efforts of several researchers, and the new methods were published by Bencivenga (1982) and Tocci (1982). These early techniques and subsequent modifications were later summarized by Alessandra Zambonelli and Rosella Di Munno (1992).

According to Zambonelli and Di Munno (1992), the early technique included the following steps. Oak seeds were first surface-sterilized by soaking in a 0.1 per cent solution of silver nitrate for 20 minutes and then rinsed thoroughly in sterile water to remove all traces of the silver nitrate, which would otherwise prevent germination of the truffle spores. The oak seeds were then either stratified in sand or used immediately. Very mature truffles were examined to ensure that they were the correct species, thoroughly scrubbed to remove all traces of soil, and rinsed in sterile water. The truffles were then placed in jars of water for a few days to hasten their natural deterioration and mimic what would normally occur in the wild. One kilogram of sugar was then dissolved in 20 L of a suspension of truffles in water to make the suspension stickier. Three oak seeds were inoculated with a sufficient volume of the suspension to contain 2 to 3 g of truffle and planted in pots containing 2 to 3 kg of natural truffle soil, which had been heated in an oven to 80°c for two to three hours to kill any competing ectomycorrhizal fungi.

Zambonelli and Di Munno also described the principal modifications to these early techniques. Instead of seeds, young, sterile seedlings raised in sterile soilless media (such as vermiculite or perlite) containing a little fertilizer were inoculated. The spore suspension was homogenized to release the spores from the small sacs (asci) they are formed in. Finally, the potting mix was autoclaved at 121°c for four hours, and the plants were inoculated under sterile conditions. There were several benefits of these modifications.

The general methods for producing truffle-infected plants. Truffle spores, truffle cultures, or truffle-infected root are used to infect cuttings or nonmycorrhizal seedlings, which are then raised in sterile soil or a specially formulated potting mix.
I. R. Hall

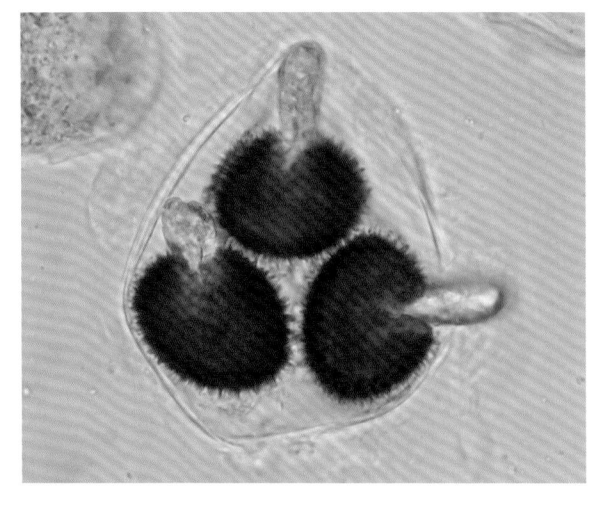

Germinating Périgord black truffle (*Tuber melanosporum*) spores C. Fischer and C. Colinas

By using seedlings, the germinating truffle spores were immediately in contact with lateral roots receptive to mycorrhizal formation. The more thorough homogenization of the inoculum made it more effective, and more attention to cleanliness ensured less contamination from other competing ectomycorrhizal fungi.

Since the early 1980s other modifications to the techniques have been developed, including the following:

- Storing truffles in sterile sand under refrigeration so that they can be purchased when truffles are cheapest and then used when it is most convenient to produce infected plants. Simply storing truffles under refrigeration also improves germination, probably through the effect of microorganisms and root exudates, and dispenses with the need to release the spores from the ascus.

- Applying a measured number of spores to each plant rather than a weight of truffle, because the number of spores per gram will vary from truffle to truffle and will be affected by the maturity of the truffles used.

- Inoculating plants with a more economical amount of truffle typically containing 10^5 to 10^7 spores—about the number contained in 1 g of truffle. Although infections can be produced by inoculating

with as few as 10^2 spores per plant, the success rate is likely to be far too low for nurseries aiming at 100 per cent to avoid the high cost of screening out uninfected plants and excluding the entry of contaminating fungi.

- Using frozen, dehydrated, or freeze-dried spores to inoculate plants.
- Sterilizing host plant seed with sodium hypochlorite or hydrogen peroxide.
- Supporting the inocula on sterile vermiculite at a ratio of 1:6 before mixing it with the potting mix to ensure even distribution of the spores.
- Replacing soil as the potting medium with modifications of soilless media used for forest plants, such as an equal mixture of peat and vermiculite; an equal mixture of peat, perlite, and limestone; or a mixture of 20 per cent rendzina or loess soil, 30 per cent peat or compost, 50 per cent vermiculite plus dolomite, to raise the pH, and coated, slow-release fertilizer granules.
- Raising inoculated plants in different types of containers, including square plastic pots, cellulose bags, black polythene bags, and moulded black polythene trays with slotted sides and open bases that allow air-pruning.
- Using containers holding less than 1 kg of substrate, so that more plants can be housed in a greenhouse.
- Heating the potting medium twice to 100°c or disinfecting the potting mix with methyl bromide in an attempt to control contaminants after inoculation.

Whatever system is used, the aim of the nurseryman must always be to infect as close as possible to 100 per cent of plants. Achieving substantially less than this would mean either screening out the uninfected plants, a very expensive exercise, or allowing poorly or uninfected plants to be planted in truffières, where they would be a sitting target for fast-growing contaminating ectomycorrhizal fungi. Consequently, those nurseries that economize by using less truffle or contaminated truffle to maximize their profit

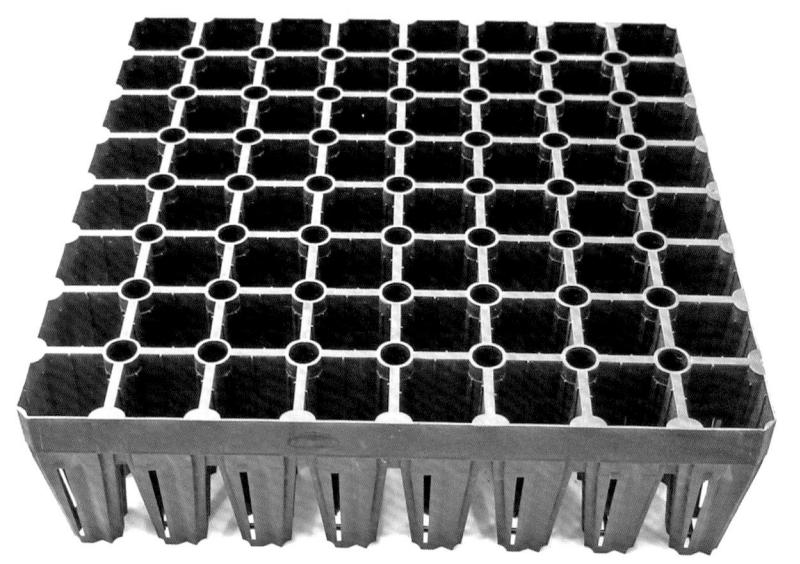

TOP Truffle-inoculated trees growing in cellulose bags. The labels indicate that the plants are heavily infected with the Périgord black truffle (*Tuber melanosporum*) and are free of contaminating ectomycorrhizal fungi. I. R. Hall

BOTTOM Lannen 64 trays with slotted sides and open bases ensure that roots are air-pruned and do not spiral in the pots. I. R. Hall

margin—or less likely so that they can sell their plants a bit cheaper—in the long run are doing their clients a grave disservice.

So far, sporal inoculation techniques have been successfully employed to produce commercial numbers of plants infected with *Tuber melanosporum, T. borchii, T. aestivum, T. brumale*, and *T. macrosporum*. This technique has also been used experimentally for *T. magnatum*, although the poor germination of its spores can cause problems.

In addition to the routine cultivation of truffles in Spain, attempts are being made by researchers in Barcelona and Valencia to inoculate naturally occurring oaks and hazelnuts in groves. The general procedure is to expose the roots using machinery, thus triggering new root growth, with or without sanitization of the soil and roots, and inoculating with relatively large quantities of sporal inocula and irrigating. Good truffle mycorrhizas have been established using these techniques, and a few truffles, but it is still early days to evaluate whether the procedures are economically justifiable.

There is no information available on whether small truffles begat small truffles or inferior, precocious truffles fruiting early in the season are likely to produce similar truffles, but it would be a sensible precaution to inoculate plants only with the best truffle available. Inoculating only with easily identifiable truffles would also limit the chances of plants being inoculated with the wrong species, a potential hazard when cheap, tiny pieces of truffle or thinly sliced dried truffle of unknown origin are used as inoculum. In addition, one large truffle weighing say 200 g is also more than 20 times faster to identify than 20 truffles each weighing 10 g.

Modifications of Talon's technique

Although spores remain the preferred method of inoculation for many nurseries, more sophisticated techniques have been developed since the 1960s. One method, the mother plant technique, is based on Talon's technique. Nonmycorrhizal seedlings are planted with infected plants in very clean, contaminant-free rooms. A similar method described by Chevalier and Grente (1973), developed by Hall in 1973 for arbuscular mycorrhizal fungi and subsequently patented by Giovannetti in 1980, dispenses with the need to have a mother plant in the same pot. Instead sections of infected

root removed from mother plants are placed close to the roots of sterile seedlings. Lo Bue and coworkers (1990) also used this method to establish truffle infections on existing trees by allowing the roots to grow through the neck of a plastic bottle containing sterilized polyacrylamide. However, the commercial value of this expensive technique has yet to be proven.

Other allied techniques simply dispense with the need to use mother plants. Instead, infected root is produced using transformed roots that are able to grow without an attached shoot, a system that has also been used for arbuscular mycorrhizal fungi. Varda Kagan-Zur and her colleagues (2002) described this technique. Derooted rockrose (*Cistus incanus*) seedlings are inoculated with *Agrobacterium rhizogenes*, a bacterium that carries the Ri (root-inducing) plasmid that induces root galls on some trees. Thin, delicate hairy roots emanating from the inoculation points of *A. rhizogenes*–treated seedlings appear 8–11 days following inoculation and are transferred individually to liquid culture medium containing antibiotics. The inoculated roots are then grown on a solid culture medium containing lower concentrations of nitrates and antibiotics, and then the cultures are transferred to a solid medium designed to support endomycorrhizal associations. The cultured roots retain viability even after several months, showing good regrowth without subculturing. A piece of agar containing actively growing Périgord black truffle hyphae is then placed in the middle of a bunch of hairy roots growing on solid medium. After about three months mycorrhizal associations develop in the elongated roots, and about two months later short clublike root forms can be observed. These techniques have since been developed to a point where they have the potential of being used commercially.

Inoculation with mycelial cultures

In 1903, Louis Matruchot became the first researcher to grow the mycelium of the Périgord black truffle in pure culture. Subsequently, Gerard Chevalier developed a method for producing truffle-infected plants with cultures or germinating spores in nutrient mist chambers. Alessandra Zambonelli and colleagues (1989) further developed the early techniques and produced standard methods for inoculating plants in test tubes using cul-

tures of bianchetto prepared from fruiting bodies. These techniques have since been used for the winter truffle, and preliminary experiments have shown that it may also be possible to produce infected plants using mycelial cultures of the Périgord black truffle and Burgundy truffle.

The patented method developed for infecting seedlings of the rockrose *Helianthemum almeriense* with cultures of the desert truffle *Terfezia claveryi* by Asuncíon Morte and Mario Honrubia (1994) is similar to those used in Italy. Plantlets of *H. almeriense* are first raised in a culture medium containing no growth hormones, under controlled temperature and light conditions. Cultures of *T. claveryi* are produced from *T. claveryi* fruiting bodies and raised on a medium adjusted to pH 8.0. The plantlets are then inoculated and further incubated until mycorrhizas form. Then they are transplanted into open pots containing a mixture of equal volumes of peat, vermiculite, and sand in a greenhouse. Similar methods have also been used to produce infections of *Mattirolomyces terfezioides* on black locust (*Robinia pseudoacacia*), which it is normally associated with, and another of the rockroses, *Helianthemum ovatum*, although the latter infections were distinctly odd.

These pure culture techniques now offer the possibility of producing truffle-infected plants free of contaminants and with strains adapted to particular combinations of soil, climate, and host. However, while these techniques are very effective for laboratory studies, more research will be needed before it will be possible to produce the quantities of mycelia needed for the large-scale production of infected plants.

THREE

Identifying Truffle Species

FALLING PRODUCTION, steady demand, and escalating international prices have been the impetus for the entrepreneur, the enterprising, and the openly dishonest to seek alternatives to the most important commercial species of truffles. It is, therefore, essential that the truffle hunter and wholesaler, the chef and the gourmet, and the producer of truffle-infected plants are able to identify truffles, have a knowledge of when species fruit and should appear on the market, and from where they originate, so that they can avoid purchasing the wrong species of truffle.

A Quick Lesson in Classification, Nomenclature, and Taxonomy

We all use common names for plants and animals in our daily life—dog, cat, rose, and truffle—but often these can be confusing. For example, what is called hazelnut in one part of the United Kingdom is called cobnut in another, *noisetier* in France, and *nocciolo* in Italy. In contrast, the word *grass* refers to many different plants, and at least one of these is likely to be smoked rather than mown. In the first part of the 18th century, Carl

Linnaeus (Carl von Linné) recognized the problems that common names pose and adopted an unambiguous way of naming organisms that had been devised more than a century earlier by Gaspard Bauhin. This was the binomial system, a system of nomenclature, which, despite newly devised molecular methods, remains the cornerstone of modern taxonomy.

The binomial system gives each distinct organism two names—a generic name and a specific name. *Homo sapiens* are the generic and specific names for modern man, *Corylus avellana* is the common hazelnut, and *Quercus robur* is the English oak. Similar species are grouped together in the same genus, so the pin oak, *Quercus palustris*, has the same generic name as the English oak but a different species name. Latin is generally used because it has fixed rules and is an extinct language, so it is unlikely to further evolve. Many Greek words are also used, as are the names of people and places, for example, *Pennantia baylisiana*, one of the world's rarest plants, is named after Prof. Geoff Baylis. To make sure that everyone knows exactly what organism is being referred to when someone names a new animal, plant, or fungus, the rules require that they also provide a full description of the organism. The originator of the name is recognized by placing his or her name or an abbreviation of it after the binomial, for example, Linnaeus or L., for the father of taxonomy.

People love to group and pigeonhole other people and things, and our languages have evolved a host of words that help us categorize—builder, doctor, politician, car salesman. Biologists and taxonomists are no different and group similar genera into families, somewhat similar families into orders, vaguely similar orders into classes, and finally classes into kingdoms. Taxonomists usually end family names with *-aceae* and order names with *-ales*. All these taxa can be arranged into phylogenetic trees to show similarities between groups of organisms. From time to time the names of newly found species and genera are added, and not infrequently the names of plants and fungi may be changed based on rules in the International Code of Botanical Nomenclature. These rules are needed to ensure the names and grouping remains useful, meaningful, and scientific. Regrettably, sometimes taxonomists lose sight of the useful and meaningful and

concentrate on the scientific, and this can create confusion in mere mortals and damage the usefulness of the system.

There are about 200 species of truffles grouped together in what seems an ever-increasing number of genera. The commercially most important genera are *Tuber, Terfezia*, and *Tirmania*, although in textbooks specializing in the taxonomy of truffles you will find names such as *Balsamia, Choiromyces, Delastria, Genea, Hydnobolites, Hydnotria, Mattirolomyces, Pachyphloeus, Picoa*, and *Stephensia*. How these genera are grouped together rather depends on which learned taxonomist you lean towards, although most mycologists recognize that the truffles are best heaped together in the order Pezizales. This is a large taxon containing, for example, the cup fungi (such as *Peziza*), morels (*Morchella*), false morels (*Helvella*), thimble fungi (*Verpa*), and many ectomycorrhizal fungi. Although this system of classification leaves out other genera that resemble truffles, like *Elaphomyces*, which are in orders other than the Pezizales, most mycologists sensibly use the common name "truffle" to encompass these as well. However, many of the fungi in other orders in the huge, diverse group of fungi called the Ascomycotina are very untrufflelike, such as the fungus that produces

LEFT Together with the truffles, *Morchella, Helvella, Verpa*, and this cup fungus (*Peziza arenaria*) are placed in the large fungal order Pezizales. I. R. Hall

BELOW Morels (*Morchella*) are related to the truffles but are very different in appearance. A. Zambonelli

LEFT False morels, such as this *Helvella crispa*, are also members of the Pezizales. A. Zambonelli

RIGHT Thimble caps (*Verpa*) are distantly related to the truffles. A. Zambonelli

ergots on grasses or the microscopic yeasts that are so important in the production of bread, wine, and beer.

Of the edible fungi, the truffles are among the most unusual. Like the cap of the button mushroom, a truffle is the reproductive organ that produces the spores that have the role of dispersing the truffle fungus. But there the resemblance ends. Truffles have no stalk, no gills, and are usually formed underground. They tend to be roughly spherical, although their shape is often moulded by stones in the soil in which they are growing. When mature, truffles tend to be firm or even hard to the touch, dense, and almost woody, rather than soft and fragile like many mushrooms.

Unlike other fungi, the truffles do not release their spores at maturity and instead have evolved strong aromas to attract consumers. These aromas are made up of complex mixtures of volatile organic compounds, including alkanes, alcohols, esters, aldehydes, ketones, and terpenes of wide-ranging polarity and molecular weight, with sulphur-containing chemicals such as

bis-methylthiomethane and dimethyl sulphide being particularly important. These attract animals, including insects and mammals, which eat the truffles. The spores then pass through the gut before being deposited in a well-fertilized piece of ground perhaps many kilometres from where the truffles were eaten. The same aromas that attract animals make some of the truffles attractive to us as well, although the disgusting aromas of others, such as *Balsamia* and *Stephensia*, make them very unlikely candidates for the table. However, the unique, pungent, and exotic aromas plus their scarcity have made some of the truffles the *prix de prix* of the edible mushrooms and have elevated them to a place in gastronomy alongside saffron, caviar, foie gras, and the finest of wines, often commanding prices beyond the budget of most.

While a few truffles, such as *Balsamia vulgaris*, are considered mildly toxic by some, their appearance and the very high prices charged for the favoured species makes it very unlikely that you would be poisoned by them—you are much more likely to suffer from the hole made in your wallet or the disappointment that comes from using an inferior species that spoils an elegant meal. Strangely, *Choiromyces meandriformis*, which is considered toxic in France and Italy for having purgative effects when eaten raw, is a delicacy in some northern European countries.

Identifying Species of Truffles

The most modern techniques used to ensure the identity of an animal or plant involve the use of the DNA present in living cells. These techniques are covered briefly at the end of this chapter. For those who don't happen to have instant access to a fully equipped molecular biology laboratory when making a truffle purchase in a marketplace or shop, knowledge of the morphological and anatomical features of the more common species of truffle may be all that lies between a superb culinary experience and disappointment.

Some specialists can distinguish species of truffle just from their distinctive aromas, but when several species of truffle are kept together for a few hours they soon take on the aroma of the most pungent. Where a

truffle was collected, what trees were growing nearby, what the soil was like, and the time of year can also be sufficient for a specialist to make an informed guess. In a marketplace, however, the origin of a truffle might not be known, and even if it is the seller might not be prepared to disclose it. It is then that more sophisticated techniques are needed.

The first thing that must be done when trying to identify a truffle is to carefully remove the soil from the surface by very gently brushing the truffle under running water, for example, with a soft toothbrush (preferably a spare one), dab it dry with a paper towel, and then carefully examine and record the external features. The truffle should then be cut in two with a sharp knife and the internal structures examined and recorded. Important characteristics include colour; surface characteristics; whether the outer surface can be easily scratched away; details of the truffle's "skin" (peridium); colour, thickness, and arrangement of the fine lines that crisscross the inside of a cut face of a truffle (gleba); and the colour and appearance of the tissue between the fine lines. Other clues can be found by using a

The quality of a truffle is in its aroma. A. Zambonelli

microscope, including the shape, size, and ornamentation of the spores and the number of spores in the small sacs that hold them (asci). These characters can then be used to identify the truffles using books, printed keys, or computerized interactive keys, such as *Truffles d'Europe et de Chine* (Riousset et al. 2001) and TuberKey (www.truffle.org/tuberkey/tuberkey-english.html), which contain descriptions, drawings, and/or photographs of the species. Permanent photographic records of the macroscopic and microscopic internal and external features of a dubious truffle are always useful. It is also important to deposit dried specimens in a herbarium, particularly if the truffles are from a commercial sample that may not be what the seller or buyer thought they were.

Périgord Black Truffle (*Tuber melanosporum*)

Although it does not fetch the extremely high prices commanded by the Italian white truffle, the Périgord black truffle (*Tuber melanosporum*) is rec-

Recently harvested Périgord black truffles (*Tuber melanosporum*). I. R. Hall

ognized by most as being *the* truffle delicacy. Its aroma and flavour are much more robust than the Italian white truffle, so it can be cooked, albeit at low temperatures, and incorporated into sophisticated recipes rather than just thinly sliced raw over hot food.

In the Northern Hemisphere *Tuber melanosporum* is harvested between late November and early March. When mature, the outer surface (peridium) of the Périgord black truffle can range from reddish brown to brown or black. It is ornamented with polygonal, four- to six-sided, slightly raised irregular warts, 2–5 mm across, with longitudinal groves and the apex depressed. The peridium, the surface of the truffle, is firmly attached to the underlying tissue, the gleba. The gleba is whitish in immature ascocarps then turns purple-black as the truffles mature and finally jet black, with thin whitish veins that can become pinkish when exposed to the air. The asci contain one to six spores, which are elliptical, dark brown, (20–) 25–55 μm by (15–) 20–35 μm (excluding ornamentation), and densely ornamented by short spines (2.5–3 μm high), which are often slightly curved at the tips.

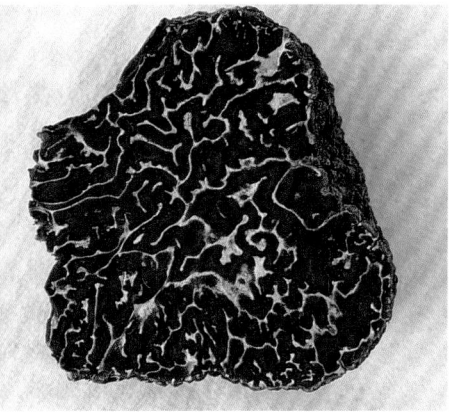 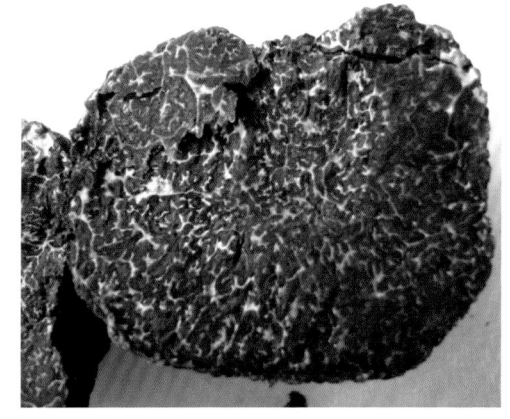

LEFT Section of a Périgord black truffle (*Tuber melanosporum*) showing the black gleba interspersed by thin, sterile white tissue I. R. Hall

RIGHT The gleba of Périgord black truffles (*Tuber melanosporum*) can turn slightly pinkish when exposed to air. I. R. Hall

Burgundy or Summer Truffle
(*Tuber aestivum = Tuber uncinatum*)

The Burgundy truffle has excellent gastronomic qualities. Thanks to its recent exposure by Italian and French chefs and researchers, *Tuber aestivum* is regaining some of the esteem it held in Victorian times. The aroma of the Burgundy truffle is intense, with a taste reminiscent of hazelnuts. It is used fresh in haute cuisine in both northern Italy and parts of France or as a substitute in recipes calling for the Périgord black truffle. It is also canned and bottled by companies such as Urbani in Italy and sold throughout the world. Some other common names for the Burgundy truffle include *la truffe de Bourgogne*, for the area of France where it is common, and *tartufo nero di Fragno* (black truffle of Fragno), for the village in the hills near Parma, Italy, where a small truffle festival is held each year. In the Northern Hemisphere the Burgundy truffle is harvested between September and late December, although it can sometimes be found as late as the end of January.

The rough surface of Burgundy or summer truffles (*Tuber aestivum*) provides the common Italian name *scorzone* (big bark). I. R. Hall

The Burgundy truffle is relatively large, with its ascocarps normally ranging from 2 cm to more than 10 cm in diameter. The peridium is brown to black, adheres to the gleba beneath, and is ornamented with five- to seven-sided pyramidal warts, 3–9 mm wide, with longitudinal fissures and some fine transverse marks. The Italian common name *scorzone* comes from the resemblance of these large warts to rough bark. When mature, the gleba is dark brown and criss-crossed with thin white veins that do not change colour when exposed to the air. The spores are covered with a reticulum formed from irregular polygonal meshes, with a mean height of 4 μm. In each ascus there are one to seven yellow-brown spores measuring 25–50 μm by 17–38 μm (excluding ornamentation).

The summer truffle, as its name suggests, is generally harvested in Europe between May and August. Its aroma and flavour are reminiscent of the Burgundy truffle but not as intense. The fruiting bodies are similar in size, shape, and colour to those of the Burgundy truffle, but the inside is hazel coloured and paler. The surface of the summer truffle also tends to have larger warts, although this is not a particularly useful diagnostic feature because of its high variability. The most important internal feature distinguishing the summer truffle from the Burgundy truffle is the height of the reticulum on the spore surface—a mean height of only 2 μm compared to 4 μm for the Burgundy truffle.

The commercial world, particularly in France and Italy, distinguish the Burgundy truffle from the poorer-flavoured summer truffle and apply the

Burgundy truffle (*Tuber aestivum*) spores are covered with a honeycomb-like network. A. Zambonelli

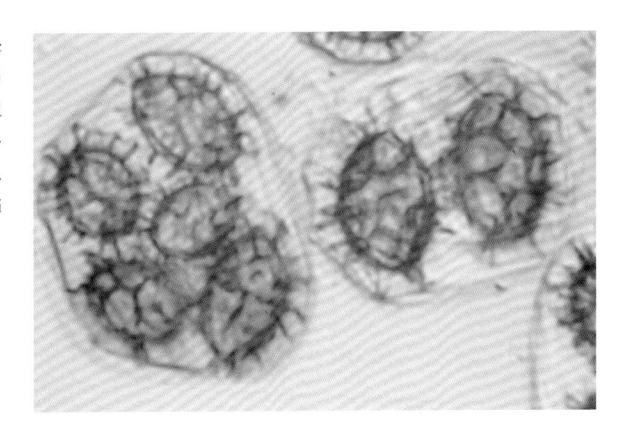

Identifying Truffle Species

name *Tuber uncinatum* to the former and *Tuber aestivum* to the latter. However, several groups of scientists have produced some strong molecular evidence to show that the Burgundy and summer truffles are a single species and that the differences between them are likely caused by environmental factors. Consequently, only the oldest botanical name, *Tuber aestivum*, should now be used.

Winter Truffle (*Tuber brumale*)

Tuber brumale is a black truffle with an odour similar to tar. Despite this, the winter truffle is purchased by large canning companies in Europe which, with the help of some flavouring agents and a special cooking procedure that drives off the tarry volatiles, produce a product that is palatable and not too different from the Périgord black truffle. Although the appearance of the winter truffle is similar to the Périgord black truffle, it is relatively easy to distinguish because the peridium of the winter truffle is only loosely attached to the gleba inside and can be peeled away with a fingernail or by brushing. Also, the warts on the surface of winter truffles are typically flat-

LEFT A winter truffle (*Tuber brumale*) showing the easily damaged surface (peridium) during washing and the broad white veins, both of which help distinguish it from the Périgord black truffle. A. Zambonelli

RIGHT A spore of the winter truffle (*Tuber brumale*) with the characteristic long, thin spines I. R. Hall

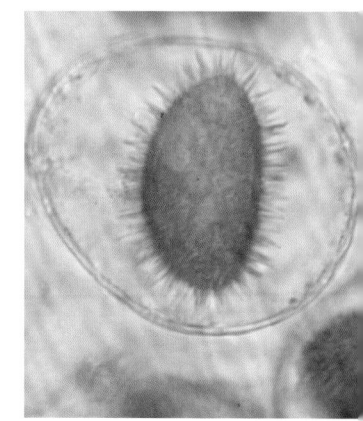

tened and depressed at the apex. There are also differences inside the winter truffle, which is greyish; never has a purplish tinge; and is criss-crossed with broad, sparse, whitish veins. The spores of the winter truffle are elliptical, smaller than those of the Périgord black truffle, (15–) 20–42 μm by 15–30 μm (excluding ornamentation), yellowish brown at maturity, and densely ornamented with straight, well-spaced, pointed spines, 3–6 μm long.

Bagnoli Truffle or Truffe Mésentérique (*Tuber mesentericum*)

The Bagnoli truffle is common in Europe, although this species is restricted to relatively small areas of France and Italy. It is particularly popular in Irpinia, Avellino, just east of Naples, and in Lorraine, France. *Tuber mesentericum* closely resembles *T. aestivum* but has an unpleasant odour of phenol and tar. Fortunately, these odours can be removed and the truffles rendered pleasant to eat if they are gently warmed or kept open to the air for a few hours.

Tuber mesentericum can be distinguished from *T. aestivum* by cutting the truffles in half to reveal the characteristic depression or cavity at the base of the truffle, with the white veins in the gleba radiating away from it. The spores are also typically ornamented with an irregular, often incomplete reticulum. Other distinguishing features are the generally smaller warts on the surface that do not have transverse markings, one to five elliptical spores in each ascus, and the slightly larger spores, (18–) 25–50 (–53) μm by (15–) 20–38 (–43) μm (excluding ornamentation).

Smooth Black Truffle (*Tuber macrosporum*)

The smooth black truffle has an excellent flavour. Although common in central Italy, this truffle is very rare elsewhere in Italy and in other countries where it has been found, namely the Czech Republic, France, Germany, Hungary, Romania, Serbia, Switzerland, Ukraine, and the United Kingdom. In Italy the smooth black truffle is found in the same areas as the Italian white truffle, fruits at the same time, and has the same host

Bagnoli truffles (*Tuber mesentericum*) with hollows in the lower surface, which are typical of this species. A. Zambonelli

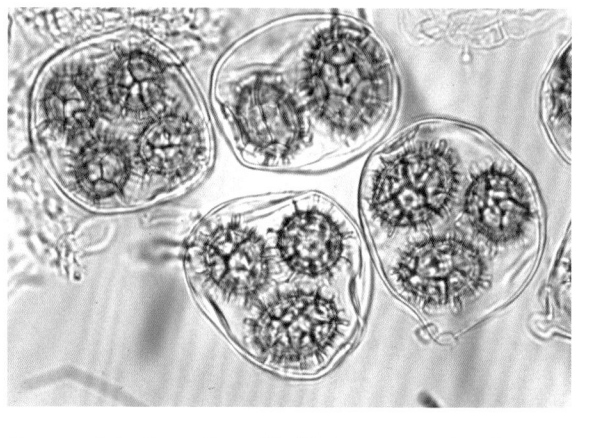

Spores of the Bagnoli truffle (*Tuber mesentericum*) are ornamented with a raised reticulum. A. Zambonelli

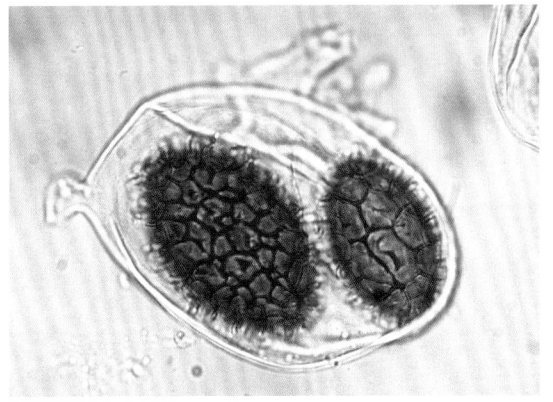

Smooth black truffles (*Tuber macrosporum*) have a surface with almost no ornamentation.
A. Zambonelli

Spores of the smooth black truffle (*Tuber macrosporum*) are ornamented with a raised, dense reticulum.
A. Zambonelli

plants—poplars, hazelnut, linden, and oaks. Generally, *Tuber macrosporum* is grossly undervalued even in Italy. Instead of being singled out as a superior species, it is sold mixed with inferior species of black truffle, such as winter, summer, and Bagnoli truffles, at truffle festivals, the usual places where truffles other than the Périgord black and Italian white truffles are sold. Consequently, with a little expertise it is possible to buy this superb species at a very low price.

The reddish brown warts on the surface of *Tuber macrosporum* are so

low and flattened that the truffle looks almost smooth. The grey-brown to purple-brown gleba is criss-crossed with thin whitish veins, which is a very distinctive characteristic, as are the dark brown, very large elliptical spores, measuring (30–) 40–80 (–92) μm by (25–) 30–55 (–62) μm (excluding ornamentation). The spores are ornamented with an irregular meshed reticulum, and only one to three spores are formed in each ascus.

Asiatic Truffles

Several truffles are traditional foods or used as tonics by the Yi and Han people of China, and the local name *wu-niang teng* (no mother plant) reflects the same perplexity Europeans had in determining the origin of truffles. One of these, *Tuber sinense*, was described in the late 1800s, but surprisingly it was not until the 1980s that truffles were "discovered" by mycologists in Chinese marketplaces. This led to the birth of new industries in the main harvesting areas of Yunnan (50 per cent the market), Sichuan, (35 per cent), and Tibet (10 per cent) and to the formal description of new species, including *Tuber formosanum*, *T. gigantosporum*, *T. pseudohimalayense*, *T. huidongense*, *T. pseudoexcavatum*, and *T. zhangdianense*.

Because some Chinese truffles, such as *Tuber indicum* and *T. formosanum*, closely resemble the Périgord black truffle, in the late 1980s a trial shipment was sent to Germany, but it was not until 1993 that regular shipments began between China and Europe. Since then Chinese truffle exports have increased dramatically. For example, in the 1994–1995 season France officially produced 13.5 tonnes of Périgord black truffle but imported 23.2 tonnes of truffles from China. The problems associated with the importation of these Asiatic truffles are discussed in Chapter 7.

Some of the Asiatic truffles that closely resemble the Périgord black truffle were originally identified as *Tuber indicum* or *T. himalayense*, while others similar to the European species *T. excavatum* (*trifola di legno*) were considered a new species. Most of the truffles resembling the Périgord black truffle and *T. excavatum* and exported from southwestern China to Europe are exported as *T. indicum* sensu lato. Although it is now illegal to import *T. indicum* into Italy due to trade restrictions, this species has been regu-

Tuber indicum, one of the Chinese truffles, has been confused with the Périgord black truffle. Y. Wang

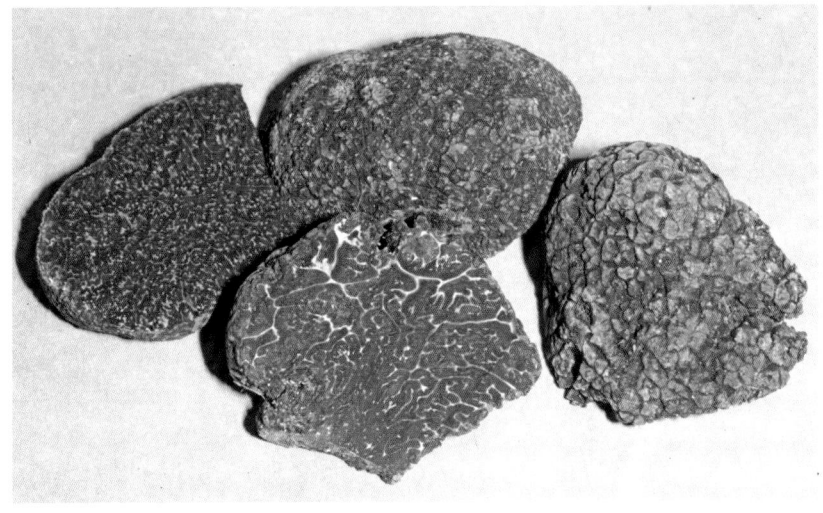

The flattened surface ornamentation of *Tuber indicum* truffles helps distinguish them from Périgord black truffles. A. Zambonelli

A scanning electron micrograph of two *Tuber indicum* spores inside part of an ascus. When mature, the spores become very dark, making details difficult to see under a light microscope. A. Pisi and A. Zambonelli

larly found in the markets there. Price, which at US$16 to US$50 per kilogram is just a fraction of the current price paid for the Périgord black truffle, has much to do with this.

Tuber indicum

Molecular comparisons of *Tuber indicum*, *T. himalayense*, *T. pseudohimalayense*, *T. sinense*, other Asiatic truffles, and the Périgord black truffle were covered in recent papers by Li-fang Zhang et al. (2005) and Yongjin Wang et al. (2006a, 2006b). These researchers made a strong case for grouping *T. indicum*, *T. himalayense*, *T. pseudohimalayense*, and *T. sinense* into just a single species using the oldest name, *T. indicum*.

Tuber indicum is the most important of the Chinese truffles because its ascocarps can be more than 10 cm in diameter and they closely resemble those of the Périgord black truffle. The fruiting bodies of *T. indicum* are highly variable, with the surface ornamentation ranging from nodulose to pyramidal warts. The peridium of *T. indicum* is 500–700 μm thick (including warts) and has two layers. The outer layer is composed of a crust of nearly globose, dark reddish black cells. The innermost layer is only slightly pigmented and is composed of intricately arranged thin-walled cells. The cells in the inner layer are paler and radially elongated in the outer zone (these are not found in Périgord black truffles) and rounded towards the inside. The gleba is dark, purplish black, and criss-crossed with numerous very thin whitish veins composed of only colourless, thin-walled cells, 5–10 μm in diameter.

The asci of *Tuber indicum* contain one to six subglobose or elliptical spores ornamented by widely spaced spines, 3–5 μm high and 1–3 μm wide at the base. Together with the ornamentation on the outside, the spores measure (15–) 22.5–25 μm by 30–35 μm. When young the spores are transparent but as they begin to mature they become reddish brown and the bases of the spines become joined, with ridges forming an incomplete net. When mature they are dark brown, opaque, and similar to those of the Périgord black truffle but a little smaller, less elliptical, and with fewer and larger elongated spines. *Tuber indicum* can be easily distinguished from the Périgord black truffle by the veins inside the truffle and its spores.

Tuber indicum is associated with *Pinus tabulaeformis* var. *yunnanensis* and *Pinus armandii* in mixed pine forests with broad-leaved trees growing in calcareous, high-pH, clay soils. This species begins fruiting when the host plants are between 10 and 40 years old and after the formation of brûlés, which can cover up to 200 m².

There are some features that distinguish the Asiatic look-alikes from the Périgord black truffle. The peridium of *Tuber indicum* is brownish chocolate coloured and lacks the reddish tones typical of immature Périgord black truffles. Also, the warts on the surface of the look-alikes are flattened, with five or six sides and with slits along the sides that are depressed in the middle and quite different from the Périgord black truffle. Despite these differences, camouflaging them with a liberal coating of soil and placing them in a closed container to concentrate the little aroma they have is often sufficient to confuse a buyer into believing that they are the real thing. Unfortunately, the chief differences between the look-alikes and the Périgord black truffle are found inside the truffle, and some of these require a good-quality microscope to see—an impossible task in a marketplace or when the truffles are inside a can or glass bottle in a shop.

Tuber pseudoexcavatum

The truffle *Tuber pseudoexcavatum* is found from August to November, 3–10 cm below the soil surface in calcareous soils under *Pinus tabulaeformis* var. *yunnanensis* at elevations ranging from 2000 to 2300 m. Its fruiting bodies are macroscopically similar to those of *T. excavatum*, but it has a brownish red peridium with flattened warts (290–500 μm thick), with a characteristic hollow at the base. The peridium has two distinct layers: an outer layer formed of subglobose, brown to reddish, thick-walled cells and a thin inner layer composed of hyaline, interwoven, thin-walled cells. The spores are elliptical to subglobose and are spiny and reticulated. The number of spores inside each ascus varies from one to eight, and the spores measure (23–) 24–28 (–35) μm by 16–19 μm. *Tuber spinoreticulatum*, described from the United States, resembles *T. pseudoexcavatum* in having spiny-reticulated spores but has leathery ascocarps and stalked asci.

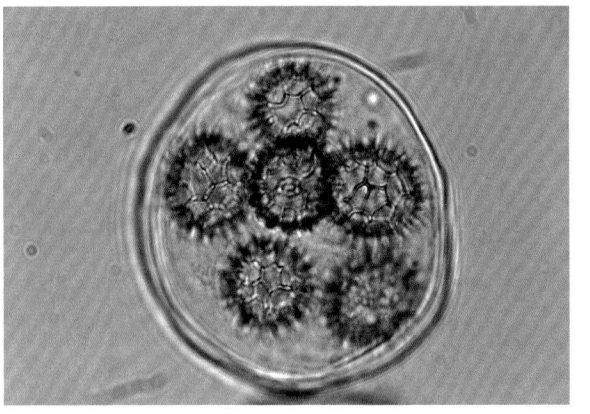

Tuber pseudoexca-vatum truffles showing the characteristic hollow in the base A. Zambonelli

The raised reticulum joining the bases of spines on the surface of *Tuber pseudoexcavatum* spores are also seen on other Asiatic species. A. Zambonelli

Italian White Truffle (*Tuber magnatum*)

The Italian white truffle commands the highest price of all truffles—usually two to five times that paid for the Périgord black truffle. *Tuber magnatum* can be found between August and January, but because of high summer temperatures and the presence of insect larvae and putrefying bacteria, truffles maturing before the middle of September have an unpleasant, almost rotten, aroma and flavour.

The excellent pungent but pleasant aroma and peculiar but superb flavour of the Italian white truffle is reminiscent of garlic and cheese but with

Italian white truffle
(*Tuber magnatum*)
J. Squires

The Italian white truffle has small spores with a raised reticulum that are not easily confused with other species.
A. Zambonelli

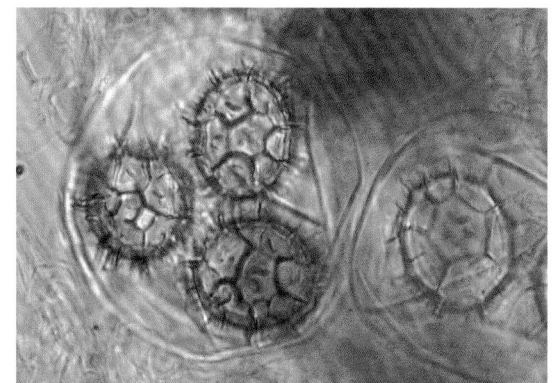

subtle undertones of methane. Chemically, the aroma is dominated by the principal volatile component *bis*-methylthiomethane (= 2,4-dithiapentane). Because this and other volatiles are lost with excess heat, the Italian white truffle is either used uncooked or added to dishes after cooking, for example, as a flavouring for pasta or salads. White truffles are also canned and bottled commercially, but the aroma is changed when the truffles are preserved and hence are inferior to those that are fresh. Cans and bottles of "white truffle" may also contain inferior species, such as *Tuber borchii, T. maculatum, T. oligospermum, Choiromyces meandriformis*, and *Tirmania*.

Fresh Italian white truffles are easy to identify. They range from just a few grams to more than a kilogram and have a smooth to suedelike surface. The colour is dependent on the habitat and host plant and ranges from pale yellowish brown to yellow-ochre, olivaceous, or greenish grey with black, rusty, or brownish spots. The peridium is composed of small subspherical cells, (3–) 9–15 µm by (3–) 6–12 (–13) µm, and the gleba is criss-crossed with numerous, fine, clear veins, which can be white to pale yellow when immature and light hazel to brown at maturity. The asci contain one to three (rarely four) round or broadly elliptical, yellowish or yellow-brown spores measuring (20–) 25–32 (–37) µm by (15–) 20–30 (–35) µm (excluding ornamentation). The spores are ornamented with a 3 to 5 µm high reticulum with wide polygonal meshes, with two to four along the greatest diameter.

Bianchetto (*Tuber borchii* = *Tuber albidum*) and Other Pale-Coloured European Species

In Italy the truffle *Tuber borchii* is commonly called *bianchetto* (whitish truffle) to distinguish it from the more expensive Italian white truffle (*T. magnatum*). Other common names are truffle of March or pine forest truffle, which refer, respectively, to the maturation time and the species' natural habitat. Although bianchetto is an excellent truffle, it has been undervalued and its use restricted to preserved products such as pâté and truffled cheese, although recently it has begun to appear on the menus of fine restaurants.

Bianchetto truffles vary from pea to egg sized and are pale yellow to red-

dish brown. The peridium is smooth or slightly pubescent, especially in the hollows. Consequently, *Tuber borchii* can resemble the Italian white truffle (*T. magnatum*) with which it is sometimes accidentally or deliberately confused. The odour of bianchetto is also similar to the Italian white truffle, although a little more garlicky. The main differences between the two species is that bianchetto is harvested during winter and early spring (versus autumn and early winter), has a darker gleba, and wider veins. However, *T. borchii* and *T. magnatum* can also be easily distinguished based on the spores. While the Italian white truffle has round spores with a reticulum with wide meshes, bianchetto spores are elliptical and have 4–6 (–10) thin polygonal meshes along the longest axis. From one to four (rarely five) spores are produced in each ascus, and their dimensions are (20–) 30–48 (–55) μm by (15–) 18–35 (–42) μm (excluding ornamentation).

Bianchetto is often confused with poor-flavoured species, such as *Tuber maculatum*, *T. dryophilum*, *T. puberulum*, and *T. oligospermum*, which are morphologically very similar. Mixtures of these truffles may smell like

When young the bianchetto truffle (*Tuber borchii*) resembles the Italian white truffle but becomes much darker as it matures. It has an excellent flavour but sells for only a fraction of that commanded by the Italian white truffle. A. Zambonelli

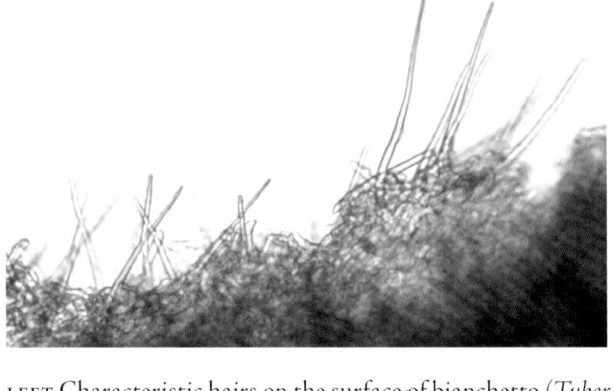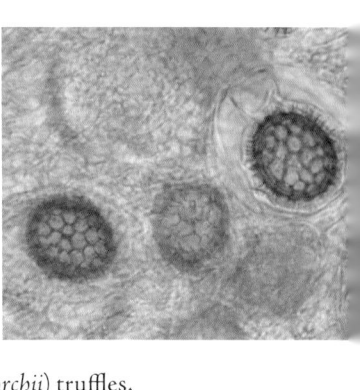

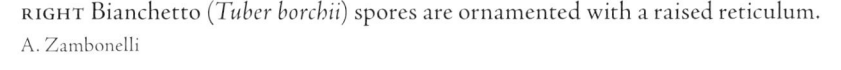

LEFT Characteristic hairs on the surface of bianchetto (*Tuber borchii*) truffles.
I. R. Hall

RIGHT Bianchetto (*Tuber borchii*) spores are ornamented with a raised reticulum.
A. Zambonelli

ABOVE LEFT These poor-flavoured *Tuber maculatum* truffles, as well as *Tuber dryophilum* and *Tuber puberulum*, reduce the value of bianchetto truffles when sold mixed together. A. Zambonelli

ABOVE *Tuber dryophilum* truffles. A. Zambonelli

LEFT *Tuber oligospermum* truffles are sometimes sold as Italian white truffles.
A. Zambonelli

T. borchii, but the flavour of food containing them is often a disappointment. Because of this, in some areas of Italy bianchetto has a poor reputation that it does not deserve. To separate these species requires careful microscopic examination of the structure of the peridium, the spore shape, and ornamentation. Particularly when young, the surface of bianchetto has prominent hairs, (39–) 41–69 (–80) μm long. The peridium is (133–) 263–547 (–713) μm thick, is predominantly composed of a mass of subspherical cells, with an outermost layer 115–300 (–399) μm thick formed of large, round cells with a diameter up to 50 μm.

In addition to the four species described below, other pale-coloured species of truffle include *Tuber asa*, *T. rapaeodorum*, and *T. scruposum*. These are primarily found in Europe and have no commercial value. Whitish truffles are easily misidentified, as evidenced by the confusion in the DNA sequences of whitish truffles submitted to GenBank. The situation is made all the more difficult because there are many other species around the world whose taxonomic validity has yet to be supported using morphological and molecular tools.

Tuber maculatum

In Europe *Tuber maculatum* is generally regarded as not fit for the table due to its slight chemical or petroleum odour and a bitter taste, although it is sold in small numbers in truffle-starved New Zealand. *Tuber maculatum* has a smooth peridium, (293–) 383–689 (–851) μm thick, composed mostly of a network of interwoven hyphae, with a thin outermost layer, 54–120 (–133) μm thick, formed by parallel hyphae and only some external roundish cells. *Tuber maculatum* spores are similar to those of *T. borchii* but have a less-dense reticulum, with only three to five (or six) meshes along the longest axis.

Tuber dryophilum

The flavour of *Tuber dryophilum* is similar to that of *T. borchii* but less intense. In Italy these two species are usually sold mixed together. *Tuber dryophilum* has a very thin peridium, (186–) 216–386 (–452) μm thick, with an external layer composed of a mass of subspherical cells, 68–156 (–199)

μm thick. The peridium is so thin that the veins of the gleba beneath can be seen. The surface of the peridium has hairs with roundish tips measuring 22–40 (–45) μm long. *Tuber dryophilum* spores are subglobose with never more than seven, usually four to six, meshes along the largest axis.

Tuber puberulum

The peridium of *Tuber puberulum* is similar to that of *T. borchii* in having prominent septate awl-shaped hairs on the surface, 60–110 μm long, but it is much thinner (150–200 μm versus about 250–550 μm). The spores of *T. puberulum* also differ from those of bianchetto, as they are round, with a dense reticulum with five to seven meshes along the longest axis.

Tuber oligospermum

Southern Europe and northern Africa are home to *Tuber oligospermum*, a truffle typical of the region. This species is quite common in Morocco and is exported to Italy and sold as Italian white truffle (*T. magnatum*).

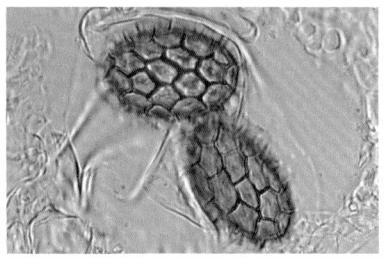
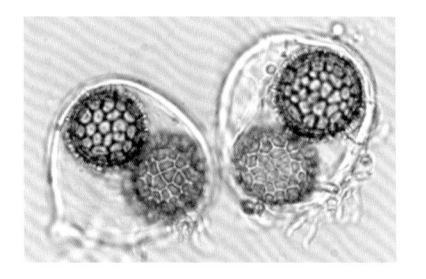

LEFT *Tuber maculatum* spores, ornamented with a raised network, loosely resemble those of the bianchetto truffle. A. Zambonelli

RIGHT *Tuber puberulum* spores are spherical, which assists in separating this species from bianchetto (*Tuber borchii*) and *Tuber maculatum*. A. Zambonelli

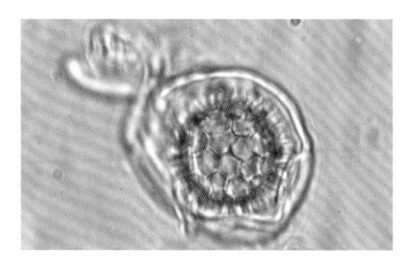

LEFT *Tuber oligospermum* spores have a smaller raised reticulum than that of the Italian white truffle (*Tuber magnatum*). A. Zambonelli

Tuber oligospermum has a very thin peridium (150–250 μm thick) composed mostly of a network of interwoven hyphae. In each ascus typically there are one to three, rarely four, spherical spores that are ornamented with a dense reticulum.

North American Pale-Coloured Truffles

Although they do not bring the high prices commanded for *Tuber mela-nosporum* and *T. magnatum*, some North American truffles have excellent culinary attributes. They are widely distributed, and industries based on their collection and sale have been established. Unfortunately, the truffles are located by raking through the soil in productive areas, rather than by using a dog, so collections often contain immature truffles and this reduces their overall value. In addition, North American truffles are often sold without prior grading, so that overall quality is lower than it might otherwise be. In addition to those described in this section, a number of Web sites, such as Bay Gourmet and Oregon White Truffles, have information on other North American truffles.

Oregon white truffles (*Tuber oregonense* and *Tuber gibbosum*)

There are at least two white truffles found in Oregon, *Tuber oregonense* and *T. gibbosum*. The species harvested in the greatest quantities is the Oregon winter white truffle; although this species has not yet been formally described, it has been called *T. oregonense* by Dr. Jim Trappe. This truffle is found from northern California to British Columbia and is only associated with Douglas fir.

The smooth surface of the Oregon winter white truffle and its colour, which ranges from whitish when young, becoming pale brown and reddish brown with age, makes it resemble the Italian white truffle. *Tuber oregonense* has an attractive strong garlicky or cheesy aroma, which becomes pungent and metallic with age, but the best Oregon winter white truffles have a sweet, musky, cedarlike aroma with hints of cinnamon, nutmeg, and vanilla. Typical prices are US$220 per kilogram.

The Oregon winter white truffle tends to be small in unmanaged forests,

but it can reach 10 cm in diameter in the cultivated soils of conifer nurseries. The gleba is white when immature, becoming brown to dark brown with age, and marbled with white veins. Each ascus contains one to three elliptical spores, 30–43 µm by 24–32 µm, which are ornamented with a double reticulum with a larger net of hexagonal meshes, 5 µm wide, and another very thin net inside with very small meshes.

The Oregon winter white truffle was not the first species of truffle found in North America. This distinction goes to the Oregon spring white truffle (*Tuber gibbosum*), which was found by the mycologist H. W. Harkness in

ABOVE Oregon winter white truffles (*Tuber oregonense*) C. Lefevre

LEFT The Oregon winter white truffle (*Tuber oregonense*; left) and Oregon spring white truffle (*Tuber gibbosum*; right) C. Lefevre

California in the late 1890s. This species has good culinary properties, with a complex odour of garlic, spices, and cheese when mature, but only now is it becoming recognized as the culinary equal of the Oregon winter white truffle. Surprisingly, *T. gibbosum* has also been reported from the markets of Reggio Emilia in Italy. The peridium is olive to brownish yellow mottled with brown patches. The surface is smooth but furrowed and contains minute hairs. When mature, the gleba is brown with white veins.

Pecan truffle (*Tuber lyoniae = Tuber texense*)

The pecan truffle was first collected in Texas in 1903 and described as *Tuber texense* by Heimsch in 1958. It is now known to occur from central Mexico to eastern New Mexico, western Texas, and Florida and from Minnesota to Quebec and Connecticut. The pecan truffle can be found from as early as June in Mexico and southern Texas to as late as March in Minnesota. Host plants for this truffle include American basswood (*Tilia americana*), scarlet oak (*Quercus coccinea*), hawthorns (*Crataegus*), shagbark hickory (*Carya ovata*), and pecan (*Carya illinoiensis*). The two species of *Carya* seem to be the most productive hosts. The peridium of *T. lyoniae* is reddish to orange-brown and smooth but creased with rough furrows. When young the gleba is white, turning brown with white veins at maturity.

    Because the pecan truffle has a good aroma and flavour, its collection and

Pecan truffles
(*Tuber lyoniae*)
C. Lefevre

Identifying Truffle Species

sale is part of the juvenile truffle industry in the United States. Research on the cultivation of *T. lyoniae* began in the late 1980s. Currently, the pecan truffle retails for about US$220 per kilogram. However, it is very important to note that poisonings can occur from pecan truffles. This is not because pecan truffles themselves are poisonous but from the very toxic pesticide aldicarb which is used in the United States to control insect damage to pecan trees.

Oregon Black Truffles
(includes *Leucangium carthusianum = Picoa carthusiana*)

Truffle harvesters in Oregon devote more time and energy to finding black truffles than either of the white truffles, probably because the black truffles command higher prices, sometimes up to US$1000 per kilogram. There are likely as many as seven species that together are marketed as Oregon black truffle, but the main one is *Leucangium carthusianum*.

In the Pacific Northwest, Oregon black truffles are found in association with Douglas fir, primarily in young, dense plantations at relatively low elevations. They are harvested from late autumn well into spring, although there are herbarium collections from almost every month of the year. The truffles are roughly spherical, typically the size of a golf ball (4 cm) but up

LEFT Oregon black truffle (*Leucangium carthusianum*) C. Lefevre

RIGHT Oregon brown truffle (*Leucangium*) C. Lefevre

to the size of a grapefruit (10 cm), have minutely warty surfaces, and are black when mature. Inside they are whitish when young, becoming grey to grey-green as the truffles mature. Although *Leucangium carthusianum* is not a member of the genus *Tuber*, its aroma has a very appealing, fruity-musky quality, but it is quite distinct from the aroma of Périgord black truffles. The words most often used to describe the high notes of the aroma are pineapple and green apple. The texture of Oregon black truffles has been described as between moist Parmesan cheese and ground almonds, with the best ones being very hard and with no sign of sponginess.

In addition to *Leucangium carthusianum*, there are other, undescribed species of *Leucangium* that occasionally turn up in the marketplace and are sold as Oregon brown truffles, but none are common. Charles Lefevre doubts if the total harvest in the best years exceeds 100 kg.

Sweet Truffle (*Mattirolomyces terfezioides*)

The truffle *Mattirolomyces terfezioides* is found from Tunisia and Sardinia in the west to possibly India in the east, but this species seems to be most abundant in the relatively dry Carpathian Basin bounded by Croatia, Slovenia, Austria, the Czech Republic, Slovakia, Ukraine, Romania, Serbia, Montenegro, and Hungary. For example, in Serbia it is common in the Deliblato Sands, located 40 km northeast of Belgrade. *Mattirolomyces terfezioides* has the flavour of a sweet cheese and is particularly popular as a food in Hungary, where it is harvested from under the black locust (*Robinia pseudoacacia*). This is odd because the black locust is not native to Hungary. In Italy *M. terfezioides* is commonly found in the littoral areas of Emilia–Romagna and Veneto in sandy soil used to grow asparagus, but it is not usually collected.

Abdulmagid Alsheikh (1994), who studied *Terfezia* for his dissertation supervised by Jim Trappe, examined specimens from India, Pakistan, and China. At that time they could not find any solid morphological basis for separating *Terfezia* truffles from *Mattirolomyces terfezioides*, although they would now like to make a molecular comparison of the east Asian collections and the European ones.

Identifying Truffle Species

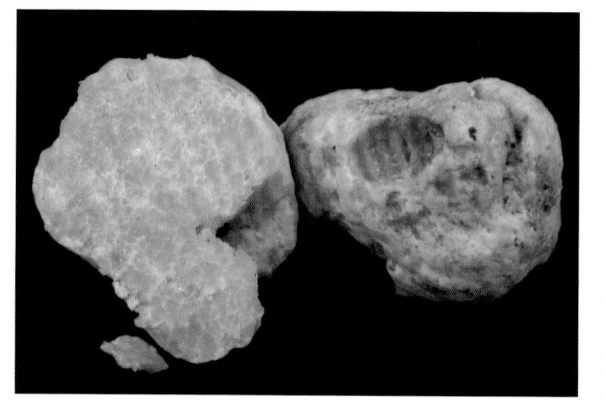

*Mattirolomyces
terfezioides*, a species
popular in Hungary
A. Zambonelli

Mattirolomyces terfezioides produces off-white to pale yellowish brown truffles between 5 and 20 cm in diameter. They look rather like the Italian white truffle, although the gleba is marbled and unlike species of *Tuber*. The asci measure about 50 μm by 125 μm and typically contain eight spores, which are about 18 μm in diameter and have a surface ornamentation somewhat similar to those of the Italian white truffle.

The only other species currently placed in the genus is *Mattirolomyces tiffanyae*. This truffle was first collected in 1998 from Story County, Iowa, in mixed deciduous upland woods containing the ectomycorrhizal species bur oak (*Quercus macrocarpa*), eastern hop hornbeam (*Ostrya virginiana*), American lime (*Tilia americana*), and white oak (*Quercus alba*). Some recent molecular work, however, suggests that this species is unrelated to *M. terfezioides*. Another species that may belong to the genus *Mattirolomyces* is found in southern Africa.

Desert Truffles (*Eremiomyces, Kalaharituber, Terfezia, Tirmania*)

The desert truffles are widely eaten in the Middle East, the Mediterranean Basin, North Africa, and as far south as the Kalahari Desert in Botswana. Desert truffles are generally regarded as a luxury, but they can also be a survival food for those who know where to find them, and David Pegler has presented a strong argument that they were the manna of the Israelites.

TOP *Terfezia arenaria* (dark-coloured) and *Tirmania nivea* (light-coloured) desert truffles
A. Zambonelli

RIGHT Kalahari truffles (*Kalaharituber pfeilii* = *Terfezia pfeilii*)
J. Dames

BELOW *Tirmania pinoyi* truffles
A. Zambonelli

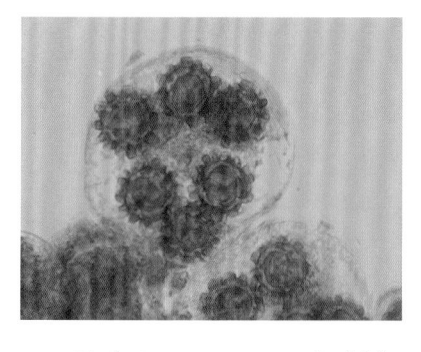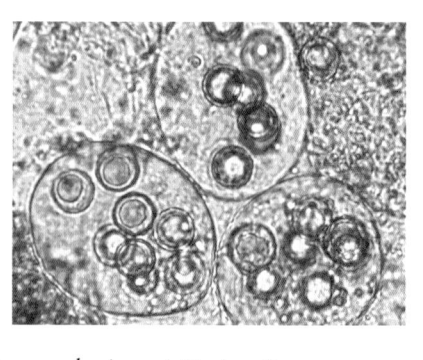

LEFT *Terfezia arenaria* spores look like tiny naval mines. A. Zambonelli

RIGHT The walls of *Tirmania pinoyi* asci and other species in the genus turn blue when treated with iodine. A. Zambonelli

Although no longer popular in Italy and elsewhere in Europe, *Tirmania* species occasionally have been exported to Europe from Iran and Syria and sold under the guise of being Italian white truffles. Desert truffles are used as a traditional medicine by Bedouin to treat diseases of the skin and eyes, diabetes, and as an aphrodisiac, and *Terfezia* are known to produce antibiotics.

The most important species of desert truffles are *Terfezia arenaria*, *Terfezia boudieri*, *Terfezia claveryi*, *Terfezia leptoderma*, *Kalaharituber pfeilii*, *Tirmania nivea*, *Tirmania pinoyi*, *Delastria rosea*, *Picoa lefebvrei*, *Picoa juniperi*, *Loculotuber gennadii*, and *Tuber oligospermum*. The common names, geographic distributions, and harvest times of these species are listed in Appendix 6. Other common names not included in the appendix include kholassi, tama, hama, and thama.

The desert truffles tend to be 10 cm or more in diameter, with the Kalahari truffle (*Kalaharituber pfeilii*) sometimes up to 40 cm across and weighing up to 1 kg. *Terfezia arenaria*, *Tirmania nivea*, and *Tirmania pinoyi* have a smooth surface, whereas *Picoa lefebvrei* is ornamented with small warts. In all desert truffles, the gleba is whitish or yellow-ochre, turning pinkish brown at maturity, and marbled from the numerous thin veins. The asci of *Tirmania* and *Picoa* contain eight smooth spores, whereas mature spores of *Terfezia* are ornamented either by truncate spines, as in *T. arenaria*, or with a reticulum, as in *T. claveryi*. A simple way of separating *Tirmania* and

Terfezia is to treat the asci with a solution of iodine. *Tirmania* asci turn blue, whereas *Terfezia* remain colourless. Recent molecular studies have erected a new genus: *Kalaharituber* for the truffle formerly known as *Terfezia pfeilii*.

Inedible and Poisonous Truffles

Choiromyces meandriformis (= *C. venosus*) is common in Europe and the United States. Although this species is considered toxic in Italy and causes mild gastrointestinal problems, it is a popular truffle in northern Europe. It can be a huge disappointment, however, if an expensive meal containing Italian white truffle was expected. *Choiromyces meandriformis* produces truffles that closely resemble those of the Italian white truffle but are generally spotted reddish brown. When immature, the aroma is weak and can be confused with that of the Italian white truffle, particularly if they have been dressed with artificial truffle oil, but when mature the aroma of *C. meandriformis* is overpowering and nauseous. The gleba has a different structure than that of *Tuber*, as it is composed of numerous fertile convoluted areolas delimited by wavy sterile bands. The yellow-brown, globose spores are also very different from those of *Tuber* because they are ornamented with truncate spines, which give the spores the appearance of miniature naval mines. In addition, there are always eight spores in each ascus of *C. meandriformis*, whereas those of Italian white truffles usually have one to three spores.

Balsamia vulgaris is the other truffle that is considered to be slightly toxic and is common in the same areas where the Italian white truffle is found. While truffle dogs usually find this species and eat it without problem, its aroma of rancid fat makes it an unlikely addition to a meal. *Balsamia vulgaris* truffles are orange-red, with the surface ornamented with small papillae, which are easily removed by brushing. The gleba is soft, whitish then turning yellowish with age, with numerous characteristic small chambers. There are nearly always eight hyaline, smooth, elliptical spores in each ascus, with a large central and two small lateral guttulae (round transparent sacs inside the spores).

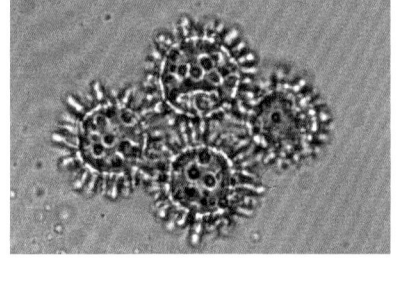

ABOVE *Choiromyces meandriformis* truffles are sometimes found mixed with Italian white truffles. A. Zambonelli

LEFT *Choiromyces meandriformis* spores are very different from those of the Italian white truffle and help distinguish the two species. A. Zambonelli

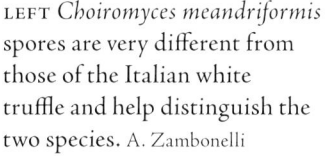

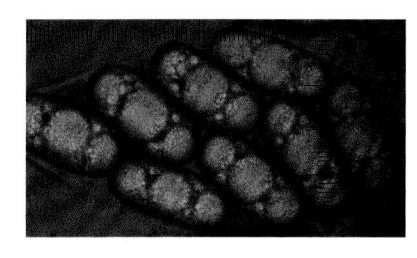

ABOVE *Balsamia vulgaris* truffles have a pale gleba. A. Zambonelli

LEFT *Balsamia vulgaris* spores are quite unlike those of *Tuber.* A. Zambonelli

Other Species of Truffle

There are many other species of truffle in Europe that are not eaten either because of their unpleasant aromas or flavours or because they are simply too small to prepare for the table. These truffles are usually found by young truffle dogs when they are looking for the most precious truffles.

Tuber excavatum (*trifola di legno*, truffle of wood) is very common in association with both coniferous and broad-leaved trees in Europe. While the aroma of this species is pleasant, it is overpowering, which restricts its use to a few hardy Italians who grate it over pasta. *Tuber excavatum* is characterized by nearly spherical ascocarps with a distinctive cavity at the base. The peridium ranges from orange to ochre to brown and is smooth or with fine papillae. The spores are yellow-brown, elliptical, and reticulated with regular meshes.

Tuber rufum is also very common in Europe, but its unpleasant acidic aroma and taste make it unfit for the table. Its ascocarps are reddish brown and usually only 1–3 cm in diameter. The surface is covered with very small warts that can only be seen with a hand lens. The gleba is the same colour as the peridium and is characterized by translucent whitish sterile veins converging at the base. The spores are ornamented with spines similar to those of the winter truffle (*Tuber brumale*).

Genea truffles are also very common in Europe, particularly Italy. At less than 1 cm in diameter and containing one or more internal chambers, they are considered too small to be worthy of the effort to prepare them for the table. Other inedible, though not poisonous, European truffles include *Tuber foetidum*, *Tuber panniferum*, *Tuber requienii*, *Tuber melançonii*, *Loculotuber gennadii*, *Pachyphloeus*, and *Stephensia bombycina*.

False Truffles

While the false truffles possess some trufflelike characters—in that they often form underground and are roughly spherical—they actually belong to a quite different group of fungi, the Basidiomycetes, a class that includes what we generally refer to as "mushrooms." During the course of their evo-

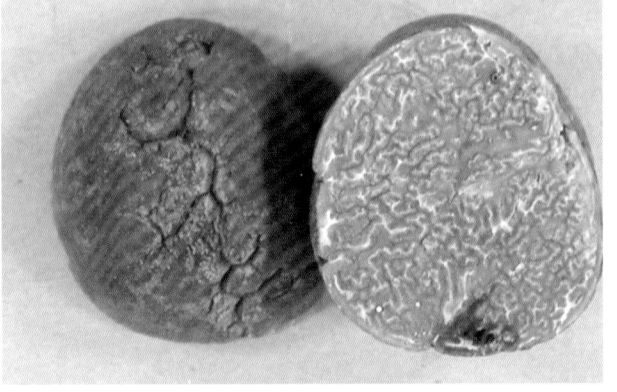

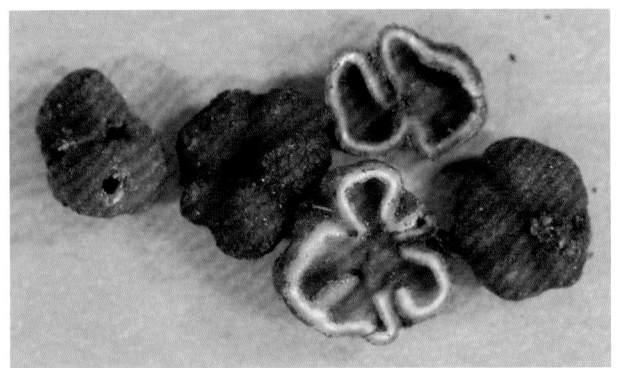

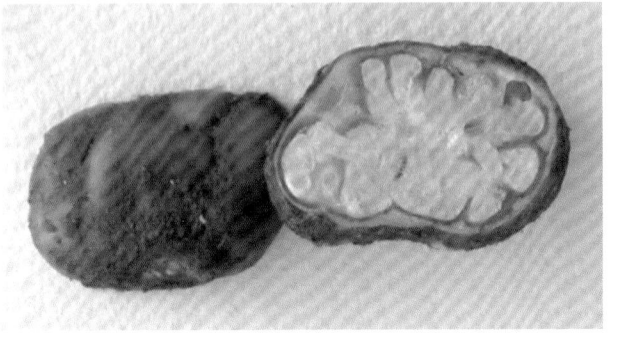

LEFT *Tuber excavatum* truffles have a characteristic hollow in the base. A. Zambonelli

BELOW LEFT *Tuber rufum* truffles A. Zambonelli

BELOW *Tuber rufum* spores ornamented with spines A. Zambonelli

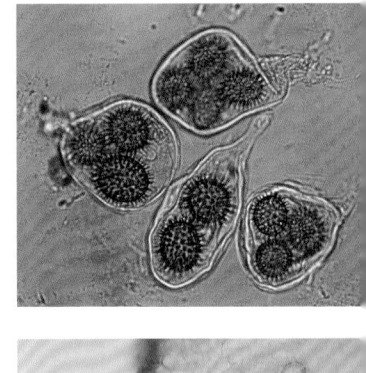

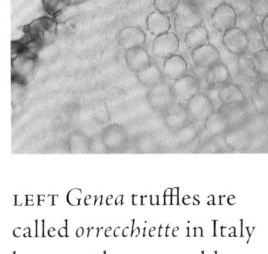

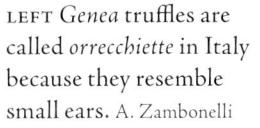

LEFT *Genea* truffles are called *orrecchiette* in Italy because they resemble small ears. A. Zambonelli

ABOVE *Genea* spores are arranged in a row within each ascus. A. Zambonelli

LEFT The foul smell and appearance of *Stephensia bombycina* truffles make them unlikely to be confused with other truffles. A. Zambonelli

lution from mushrooms, false truffles have lost the stem and the cap has been reduced to an unopened folded mass of spore-producing tissue. The internal structures of the false truffles are also quite different from the true truffles, as they lack the internal network of white lines so characteristic of the truffles and instead have a somewhat spongy interior.

There are many genera of false truffles, but some of the more common are *Gautieria, Hydnangium, Hymenogaster, Leucogaster, Melanogaster, Octavianina, Rhizopogon,* and *Sclerogaster*. A few false truffles are eaten. *Rhizopogon rubescens* (shoro) is prized in Japan, and *Mycoclelandia bulundari* is eaten by the Australian Aboriginals of Western Australia and the North-

Like all false truffles, those of the genus *Gautieria* have evolved by the gradual modification of the shape of a typical mushroom.
A. Zambonelli

LEFT The false truffle *Rhizopogon rubescens* (shoro) is considered a delicacy in Japan. I. R. Hall

RIGHT *Melanogaster ambiguus*, one of the false truffles, has an overpowering smell and is often mistaken for a true truffle by the novice. I. R. Hall

ern Territories, whereas the Arunta of central Australia believed that mushrooms were endowed with evil magic and would not eat them at all.

Identifying Truffle Species from Their Mycorrhizas

It is sometimes important to know what species of truffle are on the roots of a host plant when it is not producing truffles. One reason might be to check whether the right truffle is actually on the roots of a plant purchased from a nursery and another to see whether the right truffle has managed to survive against competing ectomycorrhizal fungi after a truffière has been established with infected plants. This can be done by first carefully taking samples of root from a plant, while making sure that the small fragile mycorrhizas are not lost in the process. The root sample is then carefully washed to remove any soil and examined first on a low-power dissecting microscope and then under a high-power microscope. The colour of mycorrhizas, size and shape of the long, thin cystidia on the mantle, and the way the cells on the surface of the mantle fit together can help to distinguish one truffle infection from another, although it takes a specialist to do so.

The yellow-brown colour of Périgord black truffle (*Tuber melanosporum*) mycorrhizas just behind the root tip, their simple or divided cystidia, and the jigsaw puzzle-like way the fungal cells fit together on the mantle surface is characteristic and sufficient to distinguish them from mycorrhizas formed by other species of *Tuber*.

Large clusters of pink bianchetto (*Tuber borchii*) mycorrhizas on *Pinus pinea* roots gathered from a 2.5-year-old truffière in New Zealand that produced at the end of its third year.
I. R. Hall

When plants are inoculated with spores of the Périgord black truffle, it is not unusual to find that spores of the winter truffle (*Tuber brumale*) have been included in the inoculum. Plants, therefore, are sometimes inspected to determine if winter truffle infections are present on the roots of nursery stock—even though this is an example of shutting the gate after the horse has bolted. Fortunately, the mycorrhizas of the winter truffle are rather different from those of the Périgord black truffle, as they have straight, unbranched, rigid, yellowish cystidia, 58–134 μm long and 3.5–6 μm wide at the base. Also, they do not branch and never ramify like the Périgord black truffle, and the mantle is formed by distinct, deeply lobed, and larger epidermoid cells than those formed by the Périgord black truffle.

The mycorrhizas of the Burgundy truffle (*Tuber aestivum*) are pale brown to brown and typically are ornamented all around the infected tips, or at least a major part of it, with long, wavy, yellow-ochre cystidia (110–600 μm high) with rounded tips, some of which are inflated. The mantle is composed of polygonal cells with rounded angles. As might be expected, summer truffle mycorrhizas are very similar to those of the Burgundy truffle. Mycorrhizas of *T. mesentericum* also closely resemble those of *T. aestivum*. The mantle is composed of a mass of polygonal cells. The tips or part of the tips of the mycorrhizas are ornamented with long, tightly packed, yellow-ochre cystidia that are sometimes dichotomously branched near the base.

Like many other species, the mycorrhizas of *Tuber macrosporum* are ochre-brown, with a mantle composed of a mass of epidermoid cells. The pale ochre, dichotomously branched cystidia are quite distinctive, however, as they are septate, 500 μm long, and slightly enlarged just under the septa.

It is very difficult to distinguish the mycorrhizas of *Tuber indicum* from those of the Périgord black truffle, which can create huge problems for someone faced with the task of determining if Périgord black truffle plants have been contaminated with this Asiatic species. However, a trained eye can distinguish the two by the slightly smaller and less lobed cells of the mantle of *T. indicum* and the dichotomous cystidia.

Mycorrhizas of the Italian white truffle are unbranched and often club shaped. They are pale to medium grey when young, turning light to

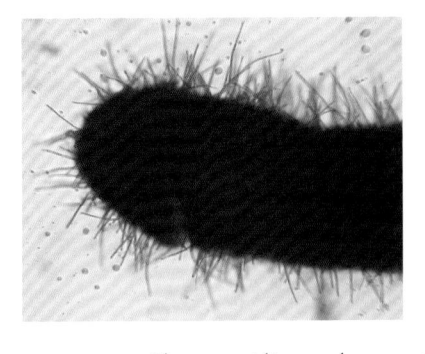
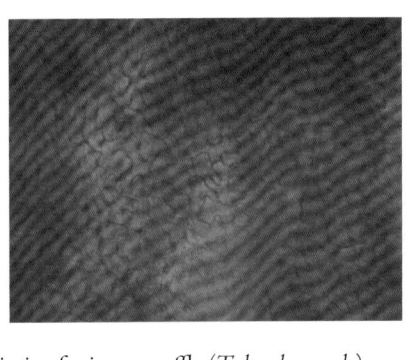

ABOVE LEFT Short cystidia are characteristic of winter truffle (*Tuber brumale*) mycorrhizas. I. R. Hall

ABOVE RIGHT The deeply lobed cells of winter truffle (*Tuber brumale*) mycorrhizas distinguish them from those formed by the Périgord black truffle. I. R. Hall

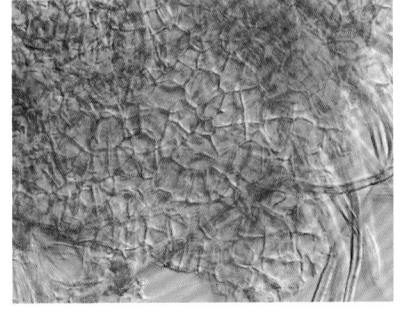
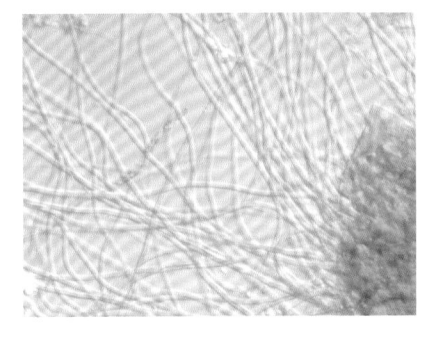

ABOVE LEFT Surface cells of a Burgundy truffle (*Tuber aestivum*) mycorrhiza
A. Zambonelli

ABOVE RIGHT The long, wavy mycorrhizal cystidia of the Burgundy truffle (*Tuber aestivum*) are very different from those of the Périgord black truffle. A. Zambonelli

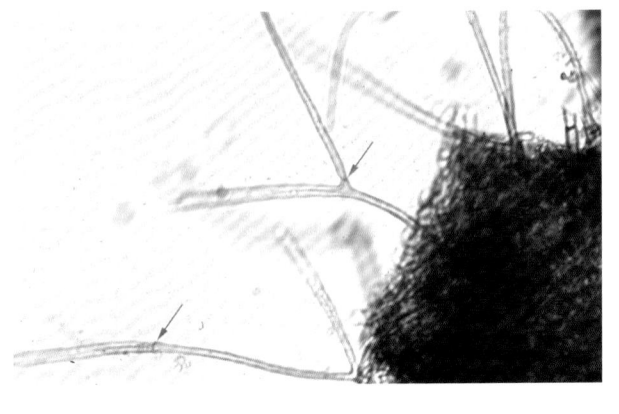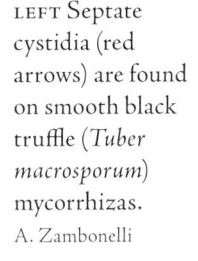

LEFT Septate cystidia (red arrows) are found on smooth black truffle (*Tuber macrosporum*) mycorrhizas.
A. Zambonelli

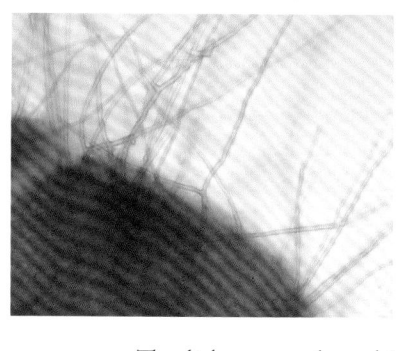 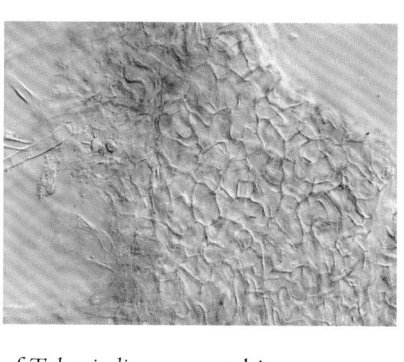

ABOVE LEFT The dichotomous branching of *Tuber indicum* mycorrhizas can sometimes be helpful in distinguishing them from those produced by the Périgord black truffle. A. Zambonelli

ABOVE RIGHT The mantle surface of a *Tuber indicum* mycorrhiza on Turkey oak (*Quercus cerris*) A. Zambonelli

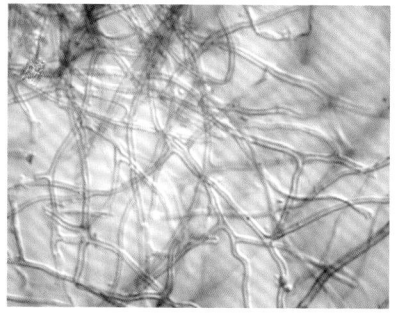 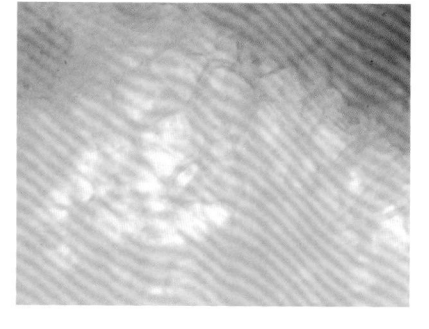

ABOVE LEFT The structure of the cystidia of an AD (*angle droix*) mycorrhiza can be confused with *Tuber melanosporum* mycorrhizas by a non-specialist. A. Zambonelli

ABOVE RIGHT The surface of an AD (*angle droix*) mycorrhiza is very different from the mycorrhizas formed by truffles but slightly similar to *Tuber aestivum*. A. Zambonelli

RIGHT *Sphaerosporella brunnea* is a common contaminant of truffle trees produced in greenhouses. The surface of the mantle is very different from that of the Périgord black truffle (*Tuber melanosporum*). The blue colour is cotton blue, a dye often used by mycologists to reveal details in colourless fungal material. A. Zambonelli

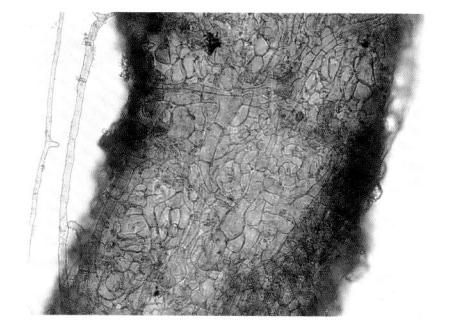

medium amber with age. Cystidia are not common but when present are up to 400 μm long and 3–5 μm wide at the surface of the mantle, have rounded tips, are simple or ramified, are arranged radially, and have multiple septa. There are also short, colourless, ramified, emanating hyphae on the surface of the mantle. In face view, the surface of the mantle has a jigsaw-like appearance. *Tuber maculatum*, *Tuber borchii*, and similar pale-coloured species of truffles (often collectively called bianchetti in Italy) have long, thin, unbranched, colourless cystidia that cover the root tips.

Some very common mycorrhizal fungi that are often found as contaminants on seedlings produced in nurseries, such as *Pulvinula constellatio* and *Sphaerosporella brunnea*, can be distinguished from *Tuber* mycorrhizas based on the structure of the mantle. For example, *S. brunnea* has brown mycorrhizas that superficially resemble those of the Périgord black truffle but are thinner and have a mantle composed of easily recognized wide, brown hyphae. The mycorrhizas of the AD fungus (from the French *angle droix*, referring to the branching of the hyphae) are superficially similar to Périgord black truffle mycorrhizas but very distinct under the microscope and unlikely to be confused by the specialist. Many basidiomycetes can be distinguished simply because their hyphae have clamp connections. Although detailed descriptions of some basidiomycete mycorrhizas can be found in textbooks, scientists now frequently rely on molecular techniques that can be used to identify a species based on only a small part of a single mycorrhizal root tip.

Identifying Truffles Using Molecular Techniques

Sometimes it is impossible to use the morphological features of truffles or spores to identify them, such as when a truffle is immature and its spores have not fully matured. In other cases, perhaps for legal reasons, there is a need to identify the species of truffle that has been used to make truffled pâté or the presence of contaminating spores in a dried spore-based truffle inoculum. It is then that new, high-tech molecular methods need to be employed. The techniques most frequently used are similar to those employed to determine paternity or identify the owner of a drop of blood

at the scene of a crime. These molecular techniques are based on the DNA differences among all species.

The first step in identifying a species using molecular techniques is to extract a small amount of DNA. The DNA is then heated to separate the double strand and then, as the DNA is allowed to cool, new copies of part or all of each strand are synthesized with the aid of a special enzyme and a complementary pair of very short stands of nucleotides, called a *primer*. The first cycle doubles the amount of DNA, the second increases it to four times, and so on until very large quantities of the DNA are produced in just about an hour. This technique is the polymerase chain reaction (PCR).

One of the first molecular methods used to identify species of truffle was the random amplified polymorphic DNA (RAPD) technique. In this method the primer is a short, arbitrary sequence of the four nucleotides (adenosine, guanine, cytosine, and thymine), which make up the DNA molecule. The primer is then used to amplify fragments of DNA. Because these fragments have an ionic charge they can be separated on the basis of their size by placing them on an agarose gel and running a current through it. This is called electrophoresis, and the resulting pattern is a fingerprint specific for a species. The RAPD technique was first selected for truffles because it requires no knowledge of the sequence of the nucleotides in the DNA of a species. Although RAPD is a powerful tool, it cannot be used to detect mycorrhizal fungi during the symbiotic phase because the short primers will also randomly hybridize with complementary sequences of DNA from the host plant. To overcome this limitation, species-specific primers for the fungi can be designed from the RAPD marker sequences.

The most useful pieces of the genetic material are the ITS1 and ITS2 spacers between the genes in the ribosomal gene cluster. This region evolves more rapidly than the genes themselves and so can be used in many ways to detect differences between species and subspecies. Some pairs of species, such as *T. brumale*–*T. melanosporum* and *T. borchii*–*T. dryophilum*, can be separated simply based on the lengths of their ITS regions. Another way of distinguishing species is to determine the nucleotide sequence for the ITS regions extracted from each species. Because each species of truffle has a different ITS sequence, it is possible to identify a species by matching its

sequence with known ITS sequences. This information has allowed scientists to develop powerful, more efficient techniques, such as specific PCR and multiplex PCR, where several different species can be identified simultaneously on one electrophoretic gel. Similar techniques may be used to determine where a truffle came from. This information might then be used to ensure that when a label on a bottle of white truffles says "Alba" the truffles do in fact originate from Alba and not from another region of Italy or even another country.

These molecular tools, which are rapidly changing the scientific world, can be used to look for overall similarities between the DNA of different organisms, to build evolutionary trees, and, importantly for the truffle specialist, to identify contaminating *Tuber indicum* or *T. brumale* mycorrhizas or their fruiting bodies in preserved food. Protocols for molecular identification of truffles can be found on the Italian-based Web site www.truffle.org.

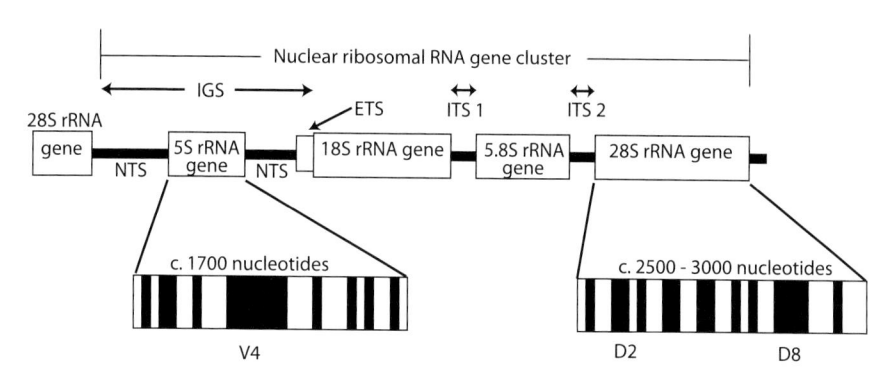

The ribosomal RNA gene cluster comprises three main genes (5.8S, 18S, and 25S or 28S) with spacer regions between (IGS, intergenic spacer; NTS, nontranscribed spacer; ETS, externally transcribed spacer; ITS, internal transcribed spacer). The spacer regions are the least conserved and evolve at a faster rate than the genes. The ITS regions evolve quickly and are useful in distinguishing between species and subspecies. Modified after Mitchell and Zuccaro (2006)

FOUR

The Habitats of Some Commercial Truffles

BEFORE ATTEMPTING to cultivate anything, whether it be animal, vegetable, or fungus, it is important to know something about the conditions of its natural habitats as well as the conditions under which it is likely to thrive. For the truffles, this includes their distribution, host trees, soil characteristics, climate, and the like. Once these things are understood, it is possible to predict where a particular species of truffle might be grown outside its natural habitat by manipulating conditions such as soil pH and soil moisture. Surprisingly, with the exception of the Périgord black truffle, Burgundy truffle, Italian white truffle, bianchetto, and one or two desert truffles, our knowledge of the ecological conditions required by all other truffle species is at best sketchy and limited to little more than lists of host plants (Appendix 3) and where they are found.

Périgord Black Truffle (*Tuber melanosporum*)

The Périgord black truffle grows naturally between the latitudes 40°N and 47°N and between 100 and 1000 m above sea level in Mediterranean, oceanic, and semi-continental to continental climate zones in France, Italy,

and Spain and less commonly in Bulgaria, Portugal, and Croatia. The typical habitats are calcareous soils on hot, exposed slopes and plateaus in marginal agricultural land. *Tuber melanosporum* is found in France in an incomplete arc from the southwest to the east of the Massif Central, the Pyrenees, and up the Loire Valley; in Italy in Abruzzo, Campania, Lazio, Liguria, Lombardy, Marche, Piedmont, Trentino–Alto Adige, Tuscany, Umbria, and Veneto; and in Spain in Alava, Cuenca, Guadalajara, Huesca, Soria, Valencia, and Zaragoza.

Between the 1920s and 1950s French collectors crossed into Aragon and Catalonia in search of truffles. The Catalonians soon learned about the prized fungus and began collecting truffles not just in their own areas but throughout Spain. From them the word soon spread to the rest of Spain. Santiago Reyna and colleagues (2005) related the words of an old Spanish truffle collector, "We saw some rare truffle hunters with their dogs looking for a type of black potatoes with a very strong smell. They had in their room plenty of full sacks. They did not want to tell us what they were doing. After two or three years we were looking for truffles." By the early 1970s most of the natural truffle areas of Spain had been identified and, armed with the research from across the Pyrenees, the first artificial Périgord black truffières were established in Spain. Truffles are now being promoted in Spain as a profitable way of adding value to forests and as a profitable crop for marginal lands.

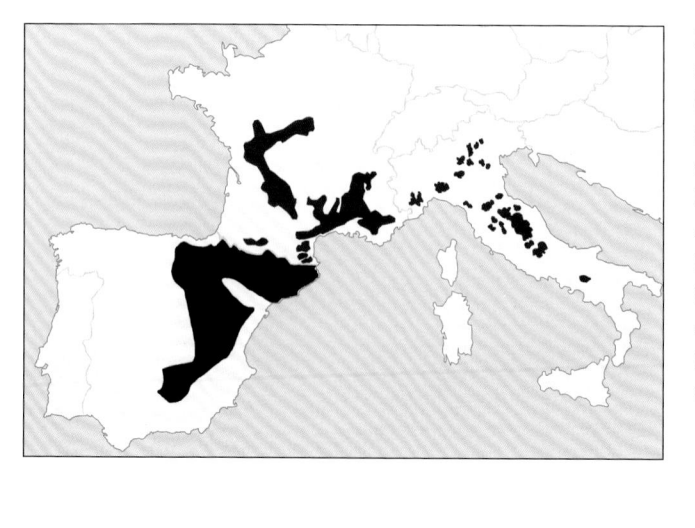

Distribution of the Périgord black truffle (*Tuber melanosporum*) in France, Italy, and Spain

Data from Delmas (1978), Zambonelli and Di Munno (1992), and Reyna et al. (2005)

While the Périgord black truffle has been harvested for hundreds of years in France and Italy, the first systematic survey of its distribution in France was not carried out until the 1960s and 1970s by Jacques Delmas, who worked in the French Institut National de la Recherche Agronomique. He determined where *Tuber melanosporum* was found, compiled maps, and defined what seemed to be the ideal conditions for fruiting:

- warm summers and cool winters (Appendix 7);
- a free-draining, high-pH (above 7.5 with an optimum of 7.9), well-aerated soil with a well-defined structure, about 400 mm deep overlying a limestone base, such as rendzinas and brown earths;
- irrigation water;
- the absence of other trees that may have competing fungi on their roots.

Similar work was subsequently conducted by other researchers in France, Italy, and Spain. However, the Périgord black truffle does not do equally well throughout the mapped areas. For example, in warm coastal Provence (the hinterland of Marseilles) *Tuber melanosporum* tends to occur on the cooler northwest-facing slopes protected from the dry southerly winds, whereas in Haute Provence (the upland, northern part of Provence) Périgord black truffles occur on the southwest-facing slopes where there is protection from the cold northerly winds.

Slope also has a bearing on the productivity of *Tuber melanosporum* truffières, with the most productive ones in France found on flat or rolling lands with a slope of less than 5°, where erosion and subsequent exposure of the delicate mycorrhizas is less likely to occur. In Umbria in central Italy, however, natural truffières occur and successful artificial truffières have been established on hills with slopes well in excess of 20°.

The general climatic parameters for the areas where the Périgord black truffle is found in France and Italy can be obtained quite simply by superimposing climatic maps, such as those provided by Arléry (1970) for France and by Cantù (1977) for Italy, over the distribution maps for *Tuber melanosporum*. For example, superimposing Arléry's mean daily July temperature map over Delmas's distribution map for France gives a mean July tempera-

TOP Scrubby *Quercus rotundifolia* are the main host plants in this unmanaged, natural Périgord black truffle (*Tuber melanosporum*) forest in southeastern Spain. I. R. Hall

BOTTOM In this managed natural truffière in southern France, the undergrowth has been removed and the trees shaped to increase the amount of sunlight striking the ground. I. R. Hall

Périgord black truffle (*Tuber melanosporum*) truffière on a steep slope in Italy I. R. Hall

ture range from about 17°C to 22°C and January range of 0°C to about 9°C for those areas with Périgord black truffles. Detailed climatic data might also be available from meteorological stations and in some cases may be downloadable straight off the Web. Another way of getting approximate climatic information for a truffle-producing area is to obtain the information for the closest city to an area producing Périgord black truffles from books and meteorological Web sites. Then allowances can be made for elevation, aspect, and exposure. For elevation this can be done by subtracting 0.65°C for every 100 m of elevation a truffière is above a nearby meteorological station, as the temperature decreases by 0.65°C for every 100-m gain in elevation. For example, the coolest part of France both in summer and winter to produce Périgord black truffles is probably Lozere which, although further south (44°N, 3°E) than many productive truffières, is cooler because it is 500

The Habitats of Some Commercial Truffles

to 2000 m above sea level. Making allowances for aspect and exposure is more difficult because they are difficult to quantify.

In areas of Europe with natural *Tuber melanosporum* truffières it rains primarily in spring and late autumn, with the summers punctuated by thunderstorms, something which has long been considered essential for good production (Chapter 1). A study of Cantù's climatic maps for Italy suggests that rainfall can be up to 1500 mm in natural truffières on the Flaminian Way near Fossombrone—somewhat wetter than the 600 to 900 mm range given by Giovannetti et al. (1994).

Périgord black truffles have been harvested in New Zealand from artificial truffières near Opotiki on the Bay of Plenty (38°s), North Island, and as far south as Ashburton on the South Island (43°s). In these areas, the ranges of mean January (summer) temperature are from 16.4°C to 20.0°C, the mean July temperature from 5.2°C to 9.2°C, annual rainfall from 729 to 1443 mm, and the number of sunshine hours from 1892 to 2397. In the United States there are productive *Tuber melanosporum* truffières 150 km northeast of Ukiah in northern California (39.5°N) and near Hillsborough, North Carolina (36°N). This species has also been cultivated in Tasmania (40–44°s) and in Manjimup in the southwestern corner of Western Australia (34.5°s). Attempts are also being made to cultivate Périgord black truffles in Chile, other countries of the Southern Hemisphere, and Israel.

From the various sources of climatic information, we have compiled the approximate climatic parameters that are present in towns near Périgord black truffle growing areas (Appendix 7). However, such comparisons do not take into account the effect of the tree canopies in truffières, the effect of the well-tended grass (a requirement around official meteorological stations), the elevation of truffières relative to the nearby meteorological station, and the cooling effects of irrigation water on soil temperatures. Consequently, various attempts have been made to obtain more accurate climatic information by monitoring the climate inside productive Italian white and Périgord black truffières. In research carried out in New Zealand beginning in the late 1990s, both air temperature and 10-cm-depth soil temperature were recorded because the truffle fungus is exposed to soil rather than air temperatures. This provided a more accurate measure-

ABOVE An ideal soil for the cultivation of the Périgord black truffle (*Tuber melanosporum*) with large quantities of limestone in the topsoil I. R. Hall

RIGHT Typical Périgord black truffle (*Tuber melanosporum*) soil profile, with up to 40 cm of limestone-rich rendzina soil over a fractured limestone base
I. R. Hall

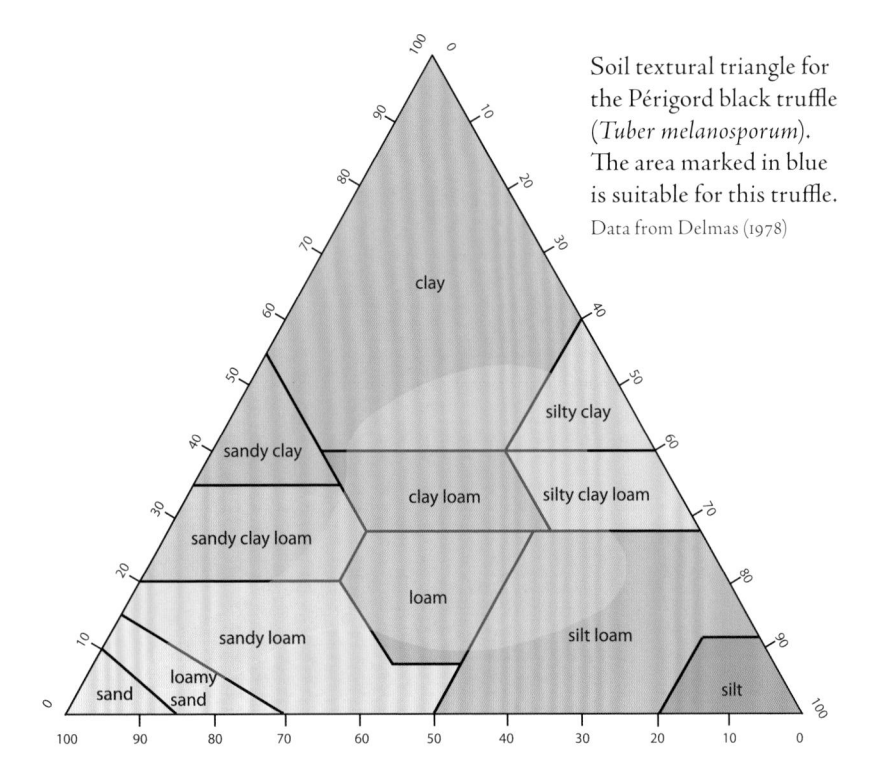

Soil textural triangle for the Périgord black truffle (*Tuber melanosporum*). The area marked in blue is suitable for this truffle.

Data from Delmas (1978)

ment of temperatures and showed, for example, that the soil temperatures in young truffières were considerably warmer than the air but fell once an essentially complete canopy had developed. Not surprisingly, in New Zealand the highest soil temperatures were recorded in the northernmost parts of the North Island and in central Otago, where low rainfall, limited ground cover, and high summer sunshine hours maximized the heating effect of the sun.

In Europe, Périgord black truffles occur naturally and are cultivated on rendzinas and brown earths. The principal characteristic of these soils is their high pH, which is caused by their considerable calcium carbonate (limestone) content. The optimum pH for *Tuber melanosporum* is 7.9, but it can be found in soils with a pH as low as 7.2 and as high as 8.3. The soils in European truffières are typically about 400 mm deep, overlie fissured limestone, and do not have an excess of clay, silt, or sand.

For some years, fruiting of the Périgord black truffle was thought to

begin in spring because truffle primordia can be seen at this time. *Tuber melanosporum* is harvested between November and March in Europe, but the truffles are probably initiated in late summer or autumn. Evidence for this came from truffle production after the very hot and dry summer of 2003, which dried the soil to a depth of more than 20 cm. Late summer rains did not eventuate in France and Italy, and the 2003–2004 season was a disaster. However, in Spain, where rain fell in late summer, production was almost normal. Had initiation occurred during spring, the primordia would have almost certainly been cooked and fruiting would not have occurred.

Burgundy or Summer Truffle (*Tuber aestivum*)

The Burgundy truffle is found throughout Europe associated with beech (*Fagus sylvatica*), birch (*Betula*), common hornbeam (*Carpinus betulus*), hop hornbeam (*Ostrya carpinifolia*, particularly important in Italy), hazelnut (*Corylus avellana*), English oak (*Quercus robur*) and holm oak (*Quercus ilex*). *Tuber aestivum* is common from Spain to eastern Europe and from Gotland, Sweden, to North Africa. In France the Burgundy truffle tends to occur in the northeast, although it is also found in smaller quantities in those areas where the Périgord black truffle dominates. In Italy the Burgundy truffle also predominates in the north of the country and is less frequent in the south. The island of Gotland, off the east coast of Sweden, is probably the coolest part of Europe where the Burgundy truffle is found, whereas centres such as Clermont-Ferrand, Paris, Perugia, and Sicily represent the warmer zones (Appendix 7).

The British Mycological Society has records for the Burgundy truffle being found in Gwynedd, Gloucestershire, Hampshire, Herefordshire, Hertfordshire, Kent, Nottinghamshire, Oxfordshire, Shropshire, Somerset, Surrey, Warwickshire, Sussex, Wilshire, and Yorkshire. It is interesting to note that the only records for Herefordshire and Shropshire are for the latter part of the 1800s. Similarly, Burgundy truffles were once plentiful in the greater Southampton area, Hampshire, and Kent, where they are never or rarely found today. However, maps showing the distribution of *Tuber*

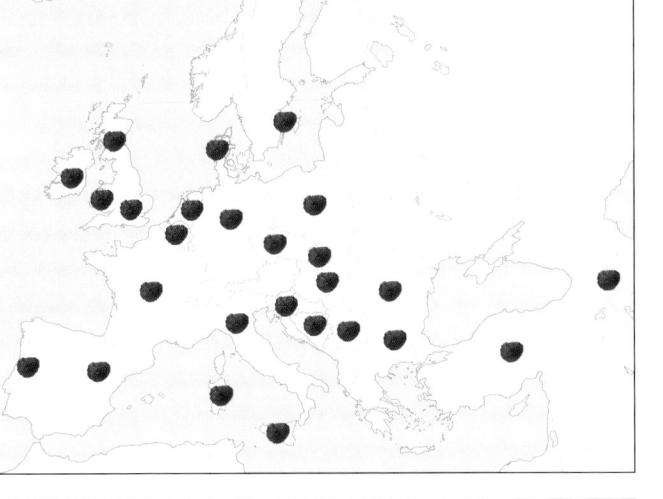

The Burgundy truffle (*Tuber aestivum*) has been found in almost all of the countries of Europe. Data from Chevalier and Frochot (1997)

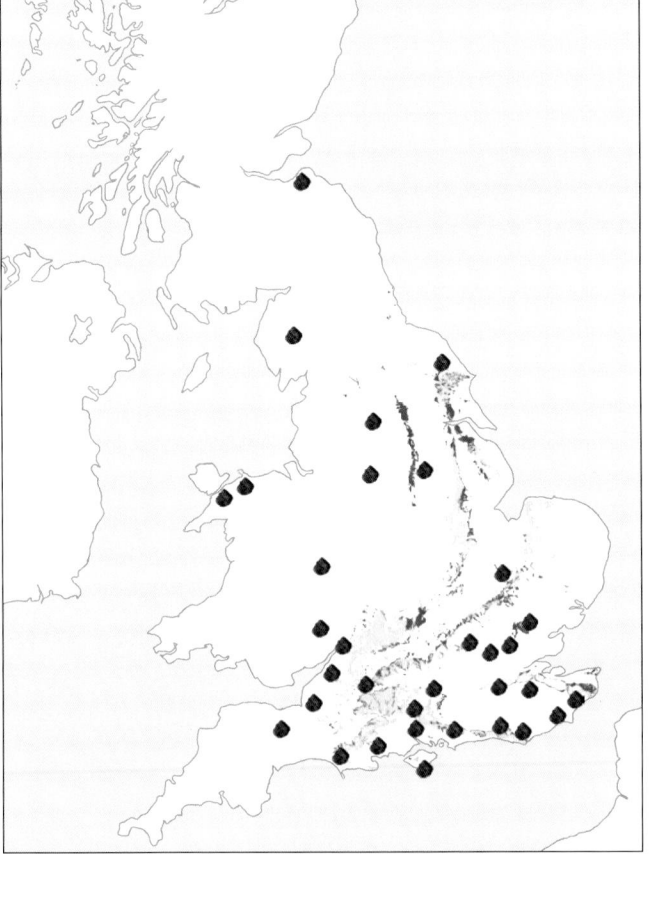

Burgundy truffle (*Tuber aestivum*) collections in England, Scotland, and Wales superimposed on maps of high-pH soils. The yellow areas are shallow, lime-rich soils over limestone, and the brown areas are free-draining, lime-rich soils. Isolated finds in Shropshire (the west midlands of England) and elsewhere represent the presence of small outcrops of limestone that are too small to feature on the map. I. R. Hall based on soil maps of England and Wales, courtesy of the National Soil Resources Institute, Cranfield University, England

aestivum within a country must be regarded with some caution because the compiler might not have been aware of unofficial finds. For example, the distribution map for the Burgundy truffle on the British Mycological Society's Web site does not show a record for the truffles found in Brighton in 1988, nor does it record the location of two large collections in 2005 of 65 kg in Somerset under holm oak and another 80 kg under European beech in Wiltshire.

The Burgundy truffle was harvested in the United Kingdom until 1937, when Alfred Collins retired. He was the last of the English professional truffle hunters and lived in Winterslow, once the centre of a thriving truffle industry in Wiltshire. Today, the Burgundy truffle is rare in the United Kingdom and when quantities are found it is sufficiently newsworthy for items to appear in the popular press. This has led to the establishment of Truffle UK, which aims to resurrect the Burgundy truffle industry in the United Kingdom.

Although *Tuber aestivum* is probably found in more countries than any other species of truffle, scientists are still learning about its distribution and ecological requirements. For example, only within the past few years has the Burgundy truffle been found growing in relative abundance on the island of Gotland. This has led Gerard Chevalier, a specialist on this species and an ardent supporter of its culinary attributes, to speculate that it may be found in more eastern European countries and may have a much wider distribution than is currently thought. A report that *T. aestivum* has been found in China, however, will need some more molecular work before this can be confirmed.

Based on detailed climatic summaries for centres adjacent to areas where the Burgundy truffle occurs, we have distilled some features that give a rough guide to the ideal climatic conditions for this truffle (Appendix 7). Compared to those areas where the Périgord black truffle is found, the natural habitats of *Tuber aestivum* occur in a much broader range of climatic conditions. For example, areas where the Burgundy truffle occurs in the United Kingdom have autumn, winter, and spring temperatures similar to Burgundy truffle areas in northern France and central Italy, but the summer temperatures are several degrees cooler than those in the south-

A natural Burgundy truffle (*Tuber aestivum*) truffière on Gotland, Sweden, showing the high level of shading, even though this is the coolest part of Europe to produce this truffle C. Wedén

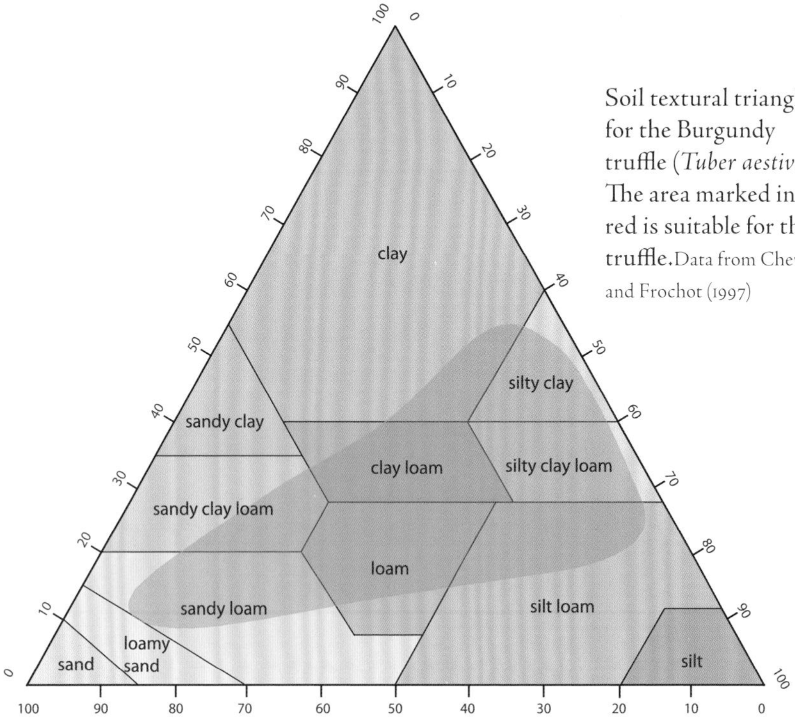

Soil textural triangle for the Burgundy truffle (*Tuber aestivum*). The area marked in red is suitable for this truffle.Data from Chevalier and Frochot (1997)

ernmost limits of the species' distribution in France and Italy (Appendix 7). Similarly, while the mean midsummer temperature in Gotland is about the same as where the Burgundy truffle grows in the United Kingdom, the mean daily temperature in winter is barely above freezing.

Gerard Chevalier and Henri Frochot (1997) listed the soil textural and chemical characteristics of 25 French soils and later put these onto a soil textural triangle, adding textural analyses from Italy and Sweden. While the Burgundy and Périgord black truffle soil textural triangles look rather similar, careful examination shows that the Burgundy truffle can be found in soils that contain more silt than Périgord black truffle soils. Burgundy truffle soils can also be quite shallow, very stony, and contain much more organic matter.

The summer truffle variety of *Tuber aestivum* is most frequently found in areas of Italy, France, and Spain with a Mediterranean climate—the southernmost parts of the species' distribution. The summer truffle is also found in the warmer Périgord black truffle areas, where it is a significant contaminating fungus in *T. melanosporum* truffières.

Italian White Truffle (*Tuber magnatum*)

Italian white truffles are found below 600 m elevation in northern Italy and up to 900 m as far south as Molise (41°N). *Tuber magnatum* is also found in Istria, Croatia, in small areas of southeastern France, Hungary, Serbia, Slovenia, and the Ticino Canton of Switzerland.

The Italian white truffle is found in woods with more or less closed canopies, areas with relatively sparse vegetation around streambeds, as well as rows or individual trees such as found near roads, near agricultural land, and in parks. There is some evidence that the distribution of *Tuber magnatum* was once wider than it is today. For example, a small area of natural forest near Bologna (Panfilia forest) on the fertile Pianura Padana produces large quantities of Italian white truffles, so it is possible that many of the fertile areas of northern Italy that have been converted to agricultural land might once have been habitats for this species.

The host plants of the Italian white truffle (Appendix 3) include Italian

TOP A natural Italian white truffle (*Tuber magnatum*) truffière in Tuscany
with a nearly closed canopy A. Zambonelli

BOTTOM Italian white truffle (*Tuber magnatum*) truffière adjacent to cultivated
agricultural land and a waterway A. Zambonelli

alder (*Alnus cordata*), hazelnut (*Corylus avellana*), hop hornbeam (*Ostrya carpinifolia*), white poplar (*Populus alba*), European trembling aspen (*P. tremula*), Lombardy poplar (*P. nigra*), Turkey oak (*Quercus cerris*), holm oak (*Q. ilex*), downy oak (*Q. pubescens*), English or common oak (*Q. robur*), white willow (*Salix alba*), pussy willow (*S. caprea*), small-leaved lime (*Tilia cordata*), and linden or large-leaved lime (*T. platyphyllos*). The most productive hosts are oaks, poplars, willows, and limes.

The Italian white truffle has broad climatic requirements. Mean January winter temperatures in adjacent meteorological stations range from 2°c to 8°c and mean July temperatures from 18°c to 26°c. The rainfall ranges from 500 to 2000 mm and is spread more or less evenly throughout the year, although in summer this tends to be in the form of thunderstorms. Because of high summer temperatures and insolation in *Tuber magnatum* areas, evapotranspiration is high, leading to subhumid climates. In some *T. magnatum* truffières inadequate rainfall is supplemented by rudimentary irrigation. In forests where little sunlight penetrates to the forest floor, air and soil temperatures in summer will be considerably lower than those that would be measured in adjacent meteorological stations. Consequently, it has been estimated that the average 10-cm-depth soil temperatures in areas where Italian white truffles grow probably do go much above 20°c.

The common feature of *Tuber magnatum* soils is an extremely soft and porous texture with pores that make up 50 per cent to more than 65 per cent of the soil volume. These pores are linked to the air above the soil, which ensures a very well-aerated structure. In addition, the soils are rich in calcium carbonate, which ensures a pH in excess of 7.5. Although there can be short periods in summer when there is inadequate soil moisture, generally the soils of the Italian white truffle are fresh and moist.

In valley bottoms where Italian white truffles are found, an open-textured soil structure develops from the chaotic deposition of mineral materials along stream banks during frequent flooding, especially in autumn but also in spring. The continuous downward movement of soil on slopes has the same effect in generating ideal conditions. Elsewhere in Italy, extensive areas of open-textured, iron- and manganese-rich soils that have been developed from Miocene marls through frequent tillage are also ideal

Moist valleys like this one near Savigno, Emilia–Romagna, are typical habitats for the Italian white truffle (*Tuber magnatum*). Note the primitive irrigation system used to boost production.
A. Zambonelli

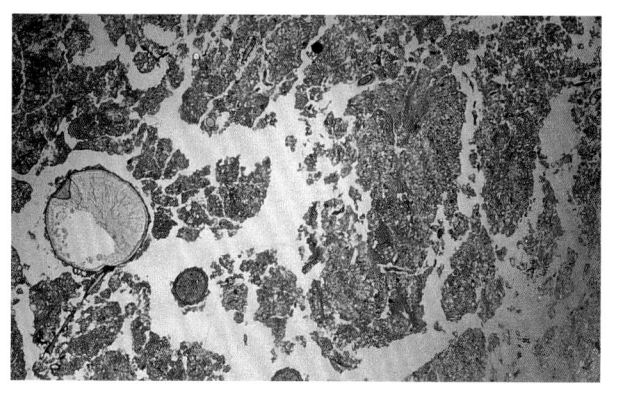

A section of an Italian white truffle (*Tuber magnatum*) soil showing the large (white) air spaces
M. Pagliai and F. Primavera

for the Italian white truffle. *Tuber magnatum* is often found in small areas where these materials have been trapped, for example, by fallen branches and tree trunks. In each of these situations, the open texture of the soils is assisted by frequent freeze-thaw cycles in winter and wetting-drying cycles in summer. Italian white truffles are only rarely found in streambeds where the deposition of fine material has filled up the soil pores. The highly aerated nature of *T. magnatum* soils also ensures that immediately after one of the sporadic but heavy thunderstorms that punctuate summer in Italy and southern France, the air within the soil is rapidly replaced by the heavier, cooler, oxygen-rich air from above the soil. This sudden change in the soil temperature and air composition could be the biological trigger for truffle formation. Yields are highest when a moist autumn is followed by a winter with many freeze-thaw cycles and a summer with many thunderstorms, with soil temperatures not exceeding 20°c.

Analyses of Italian white truffle soils have been conducted, but differences in extraction procedures and omitted nutrient tests make it difficult to equate these with analyses from elsewhere. Consequently, in the mid-1990s the authors analyzed samples collected from Italian white truffle soils using standard New Zealand techniques (Appendix 8). The high concentration of calcium carbonate in these soils ensures that any carbon dioxide produced by biological activity is rapidly removed by calcium ions and ensures that the ratio of oxygen to carbon dioxide is always high.

Infections even in productive parts of natural *Tuber magnatum* truffières tend to be very sparse, typically occupying about 3 per cent of the root tips—about the levels of infection that can be obtained using standard inoculation procedures. These sparse infections seem unlikely to be able to support the quantities of truffles that are harvested, so it has been suggested that we are missing a vital piece of the puzzle that is preventing us from cultivating this most prized species of the fungal world. One suggestion is that a third organism may be needed to complete the habitat for the Italian white truffle and some other edible ectomycorrhizal fungi to enable them to grow and fruit. Bacteria seem to be a possibility, as are other ectomycorrhizal fungi.

Because the cultivation of the Italian white truffle has not yet been mas-

tered, considerable attention is now being paid to maintaining conditions in natural truffières in an attempt to ensure continued or even enhanced production. This has led to detailed studies of the ecology of *Tuber magnatum*, including the effect of canopy density, soil aeration, and soil moisture on production. Based on such ecological findings, an apparent breakthrough in the cultivation of this most highly prized fungus was announced at a small conference in Finland late in 2006. The discovery that *T. magnatum* reproduces sexually and that most of its life cycle is spent in the haploid state (that is, containing half the normal number of chromosomes) may have practical importance in producing mycorrhizal plants for use in artificial truffières.

Bianchetto (*Tuber borchii*)

Bianchetto is found throughout Europe, from southern Finland to Sicily and from Ireland to Hungary and Poland. This truffle, however, is said to be rare in England, Wales, Ireland, Denmark, Switzerland, and Germany. It is found almost throughout Italy, in Abruzzo, Basilicata, Calabria, Campania, Emilia–Romagna, Friuli–Venezia Giulia, Lazio, Liguria, Lombardy, Marche, Molise, Piedmont, Puglia, Sardinia, Sicily, Trentino–Alto Adige, Tuscany, Umbria, Valle d'Aosta, and Veneto. Although *Tuber borchii* has been reported from China, the molecular techniques used to confirm its presence were probably inadequate to separate bianchetto from other species it can be confused with. (Further work is needed on bianchetto DNA sequences deposited in public databases, some of which are clearly incorrect and a consequence of incorrect morphological identifications prior to molecular characterization.)

Bianchetto is commonly found in calcareous, very sandy soils, such as those in coastal areas of Italy, as well as in the same alkaline soils where the Périgord black and Italian white truffles are harvested in the Apennines. In Italy it is also found in neutral and slightly acidic soils where the pH ranges from 6 to 7, although occasionally the pH can be as low as 5.2. The climate of areas where bianchetto is found ranges from cool temperate to Mediterranean, with rainfall ranging from 600 to 1600 mm. The climates

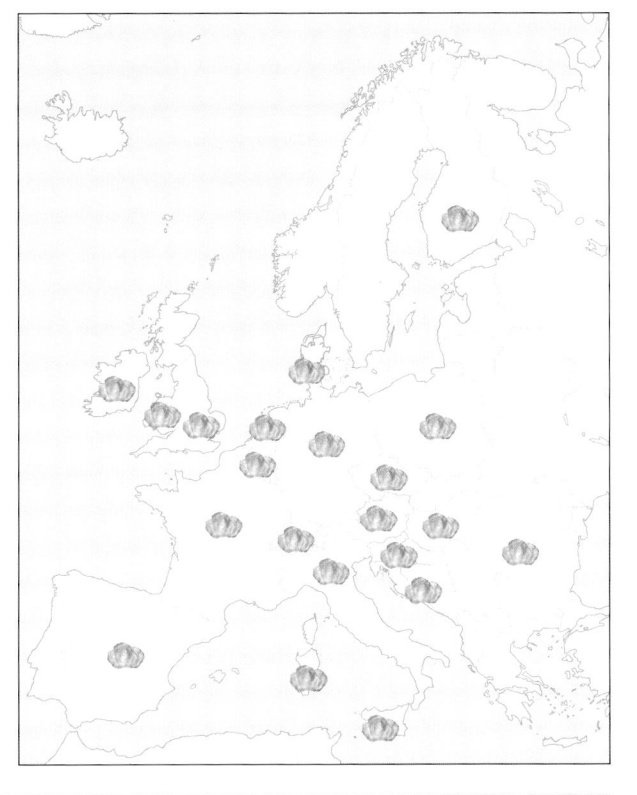

RIGHT Areas in Europe in which the bianchetto truffle (*Tuber borchii*) can be found I. R. Hall

BELOW A typical natural bianchetto (*Tuber borchii*) truffière in the Apennines I. R. Hall

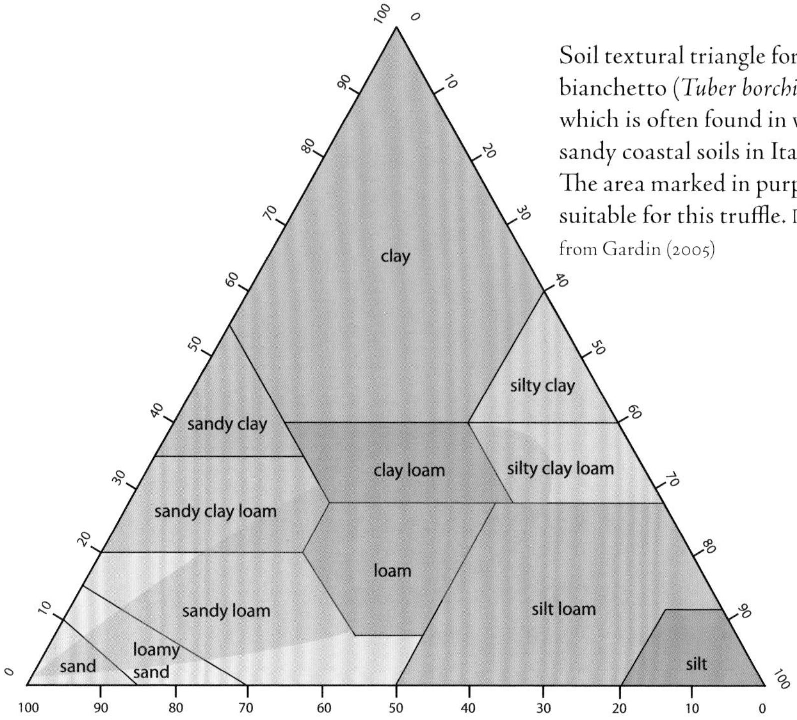

Soil textural triangle for bianchetto (*Tuber borchii*), which is often found in very sandy coastal soils in Italy. The area marked in purple is suitable for this truffle. Data from Gardin (2005)

Bianchetto (*Tuber borchii*) is common in coastal sandy soils under stone pines (*Pinus pinea*), as in this natural truffière near Marina di Ravenna, Italy. I. R. Hall

of Helsinki, Finland; Durham and Yeovilton, United Kingdom; Wroclaw, Poland; and Cervia, Perugia, and Palermo, Italy, probably span the climatic extremes where *Tuber borchii* is found (Appendix 7). The wide ecological range of *T. borchii* suggests that it should be possible to cultivate this truffle in those areas with climates and soils unsuited to the cultivation of the Périgord black or Italian white truffles.

While the principal host plants for bianchetto in Italy are the stone pine (*Pinus pinea*) and maritime pine (*Pinus pinaster*), it is also found associated with a wide range of other species, including cedars (*Cedrus*) and numerous broad-leaved trees (Appendix 3). In France bianchetto is found particularly in association with oaks (*Quercus*) and Turkish pine (*Pinus halepensis*) and in the United Kingdom with larch (*Larix*) and beech (*Fagus*).

Desert Truffles (*Eremiomyces, Kalaharituber, Terfezia, Tirmania*)

The desert truffles are found in the Middle East, throughout the drier parts of Africa as far south as the Kalahari Desert, and in the drier areas of France, Sardinia, Sicily, and Spain (Appendix 6). As their common name suggests, the desert truffles inhabit arid regions with sandy soils. Although the soils are very free draining, excessive dryness can damage *Terfezia* production, as can freezing. *Terfezia boudieri*, *Terfezia claveryi*, and *Terfezia olbiensis* inhabit alkaline soils with pH ranging from 7.4 to 8.5, whereas *Terfezia arenaria* and *Terfezia leptoderma* are found in mildly acidic soils with pH less than 7.0. In the areas where Mario Honrubia and colleagues are producing large quantities of *T. claveryi* truffles in Spain, the soil texture ranges from 18 to 53 per cent sand, 27 to 43 per cent silt, and 28 to 39 per cent clay, and there is a very high proportion of active calcium carbonate. Most of the host plants of desert truffles are in the rock rose family (Cistaceae) and the genera *Cistus* and *Helianthemum*, but the camel thorn (*Acacia erioloba*) is also an important host in southern Africa (Appendix 6).

Truffle fungi and their host plants form typical ectomycorrhizas, but the helianthemoid or terfezoid mycorrhizas formed by *Terfezia* and *Tirmania* are very different. The mantle, when formed, is a loosely arranged hyphal

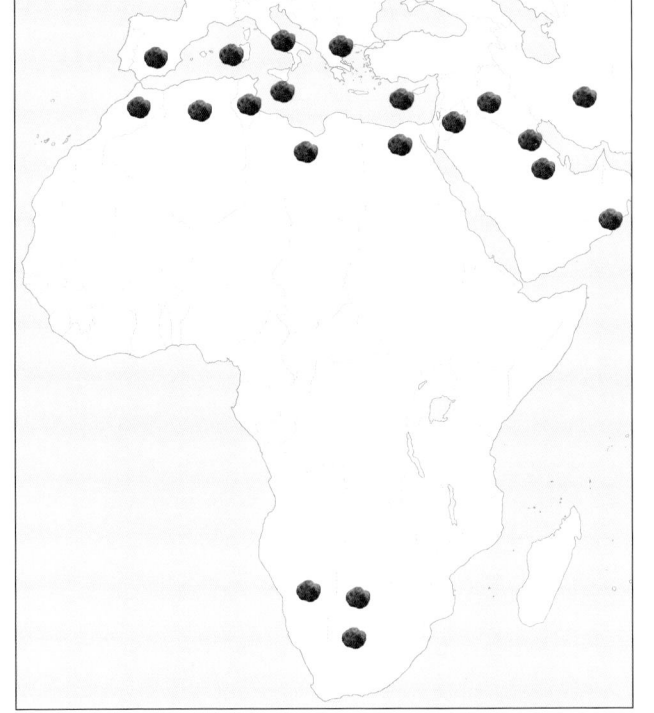

Areas in
Europe, Africa,
and the Middle
East where
Terfezia and
Tirmania
desert truffles
can be found
I. R. Hall

envelope; a Hartig net, if present, can be rudimentary; and hyphal coils can fill the cortical cells in the outer layers of roots. Just by changing the growing conditions, in particular the phosphorus concentration of the media, researchers have found that they can change the structure and appearance of desert truffle mycorrhizas. Some other plants, such as watermelon (*Citrullus vulgaris*), can also be infected by desert truffles, but whether the relationship between the plant and fungus is a mycorrhizal one, with both partners in the symbiosis benefiting from the relationship, is debatable. In many ways the infections found on the watermelon are rather reminiscent of the anomalous Périgord black truffle infections Isabel Plattner and Ian Hall (1995) found in grass roots growing around trees in a truffière and which they considered to be pathogenic and a probable cause of the brûlé. Interestingly, people of the Kalahari Desert expect to find *Kalaharituber pfeilii* under grasses, particularly on the southern-facing, shaded sides of camel thorn.

Like other commercial species of truffle, desert truffles are threatened in

some parts of the world. They were once common in the Muqattam Hills to the east of Old Cairo, but urban development has all but destroyed production in this area. The same problem is also threatening production in the United Arab Emirates, while the Olaya district of Riyadh, an area that produced truffles in the 1970s, is now almost entirely under concrete. Areas of Kuwait that produced desert truffles also suffered during the Gulf War, almost certainly due to the ground being stirred up by heavy vehicles, the subsequent erosion of the fragile desert environment, and loss of host plants. Chemical pollution is also a likely cause, with at least 25 per cent of Kuwait's desert having been covered with oil or heavy deposits of acidic, oily soot during the war. In Morocco the numbers of Beni Guil pastoral nomads who harvest desert truffles has gradually shrunk as they move away from a nomadic lifestyle to live in permanent settlements.

Because of the perceived fall in truffle production in Saudi Arabia, there have been recommendations by the Saudi Arabian Environmental Protection Agency that collecting should be regulated and collectors licensed. These regulations have been hotly debated, however, with some arguing such regulations are unlikely to be heeded by the free-thinking Bedouin. Desert truffle harvests have also declined in parts of the Kalahari, which has been attributed to soil disturbance through overstocking with cattle and goats, although this is also disputed by some.

While the traditional belief that the size of desert truffle harvests is linked to thunder is rather unlikely, fruiting is related to sufficient rain falling at the right time. In the Middle East a minimum of 180 mm of rain needs to fall between October and March. Research suggests that the optimum conditions are 200 to 250 mm of rain between October and December (Northern Hemisphere) followed by a relatively dry period in January and then light rain between February to April, during the harvest. Whereas excess rain can result in the young truffles decaying and a failure of the harvest, in artificial *Terfezia claveryi* truffières in Spain increasing the equivalent rainfall from 250 to 350 mm/year was shown to produce a 5- to 10-fold increase in yields.

FIVE

Establishing a Truffière

IN EUROPE, PARTICULARLY Italy, those who live in country districts are well aware of the species of truffle found in their area. This knowledge is reinforced with information in books and Web sites that detail where each truffle is found, the most suitable soils, host plants, and the like. Soil may be tested to ensure that the soil chemistry and texture are suitable before establishing a truffière, but generally the overriding factor in the decision-making process is that the truffle of choice is known to grow or have grown in the area. The probable presence of competing ectomycorrhizal fungi, such as the winter truffle in Périgord black truffières in France, summer truffles in Périgord black truffières in Italy, or *Tuber maculatum* in Italian white truffières, are viewed almost as inevitable and lamentable problems to be managed to minimize the impact.

Outside of the normal range where truffles occur naturally, and particularly beyond the confines of Europe, a different approach to truffle cultivation has been taken. In these areas there is no traditional knowledge, so it has been necessary to follow a more fundamental approach. Before establishing a truffière, one must determine whether the climate is within the

species' natural range; whether the soil is suitable and, if not, whether it is possible to modify it; and how competition from contaminating ectomycorrhizal fungi can be limited rather than tolerated. Answering such questions requires an in-depth knowledge of the ecologies of each of the commercial truffles (Chapter 4). Regrettably such information is still rather fragmented, but some basic principles are understood.

Competing Ectomycorrhizal Fungi

Scraping away the litter layer under an ectomycorrhizal tree will reveal a mass of fungal threads covering the decaying leaves. Many of these will be the ectomycorrhizal fungi on which the tree depends, and it is these fungi that are the potential competitors for space on a truffle-infected plant. Studies have found up to 600 m of ectomycorrhizal fungal hyphae in each gram of soil, which arguably makes this the most important threat facing a truffle grower. Fortunately, not all of the 5000 to 6000 ectomycorrhizal fungi are a problem because they are unable to infect truffle host plants. For example, a survey of more than 200 ectomycorrhizal fungi in the United Kingdom showed that about 40 per cent were associated with just a single host genus. Similarly, eucalypt-compatible ectomycorrhizal fungi rarely form mycorrhizas with fungi originating outside of Australia, and in Tasmania very few Australian native ectomycorrhizal species infect truffle oaks.

Specificity extends beyond the host-fungus relationship, with some ectomycorrhizal fungi more successful in one location than another. Adaptation to soil physical conditions, such as temperature and moisture, may be one factor affecting success, while having a suitable suite of enzymes that can break down and access components such as chitins, cellulose, proteins, and lignin present in particular types of soil may be another. Nevertheless, there are some ectomycorrhizal fungi, including *Hebeloma*, *Inocybe*, *Scleroderma*, and *Tomentella*, that are able to thrive in high-pH soils and form stable mycorrhizas with truffle host plants, thus having the potential to cause major problems in a truffière.

TOP The white, cottony mycorrhizas of the genus *Scleroderma* have a characteristic smell and are common mycorrhizal contaminants in truffières. I. R. Hall

BOTTOM *Scleroderma* fruiting around the base of a tree infected with Périgord black truffle (*Tuber melanosporum*) I. R. Hall

Choosing a Site for a Truffière

Selecting the ideal location to cultivate a truffle or choosing a truffle for a particular location comes down to climate, aspect, canopy coverage, and soil characteristics, including moisture, texture, chemistry, and pH (Chapter 4). With care, competition from ectomycorrhizal fungi should be minimized and shifted in favour of the truffle.

Because of the dangers competing ectomycorrhizal fungi present, the ideal place to establish a truffière is one where the land is completely dominated by arbuscular mycorrhizal plants—the vast majority of cultivated crops, flowering plants, ferns, cycads, and some gymnosperms—and ectomycorrhizal plants are absent. Areas that have been used to grow crops, grass and clover pastures, and vineyards are possibilities. Indeed, the lack of competing ectomycorrhizal fungi in vineyards is a probable reason why the truffières established in the *Phylloxera*-ravaged vineyards of France in the second half of the 19th century were so successful and led to the heyday of the truffle industry (Chapter 1). However, the simple absence of an ectomycorrhizal host tree in an area is not sufficient to conclude that ectomycorrhizal fungi will be absent. It is likely that the vegetative stage of an ectomycorrhizal fungus (the hyphae in the soil and on and in the root) will survive for a few years after a tree has died or been felled and may spread onto the roots of a truffle-infected tree planted in the soil. Similarly, long-lived spores and resistant perennating structures produced by ectomycorrhizal fungi that once occupied a soil may still be present long after all signs of the tree have disappeared. Attempting to remove the roots of ectomycorrhizal trees after felling is unlikely to be successful because it is impossible to remove all the small fragments of infected root and the tiny resistant fungal spores.

In Europe, truffières are often established where there are contaminating ectomycorrhizal fungi or even in newly cleared scrub or forest that had contained ectomycorrhizal trees. In these situations, the only way to diminish the negative impact of competing fungi is to choose an area ideally suited to the particular truffle that is to be cultivated so that it is able to survive and outcompete other ectomycorrhizal fungi. To minimize competition from

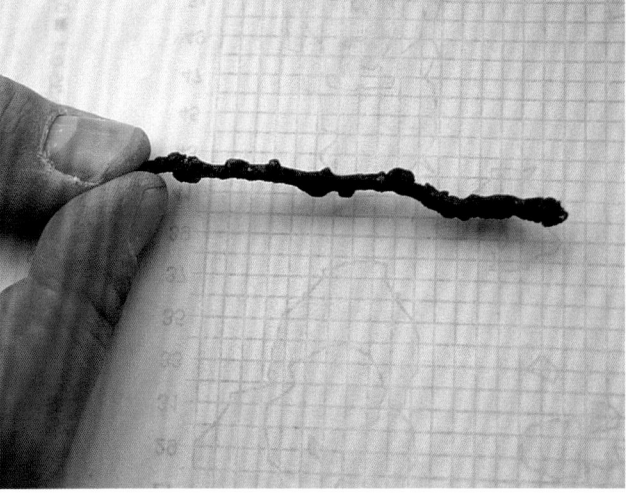
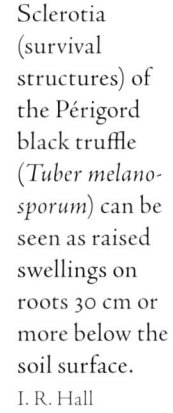

Sclerotia (survival structures) of the Périgord black truffle (*Tuber melanosporum*) can be seen as raised swellings on roots 30 cm or more below the soil surface. I. R. Hall

In Europe these reddish iron-rich soils are considered particularly well suited for truffle cultivation. I. R. Hall

other ectomycorrhizal fungi, however, preparation of the ground often begins several years before planting, with the removal of old trees, ripping out the root systems with a bulldozer, and then growing crops with arbuscular mycorrhizas in the ground for several years. In France, treating the soil with methyl bromide or 3,5-dimethyl-1,3,5-thiadiazinane-2-thione has only been partly successful in eliminating competing fungi. However, the use of methyl bromide is now restricted because it contributes to depletion of the ozone layer.

Unfortunately, even in locations without any competing fungi, such as large areas of New Zealand, sooner or later tiny airborne spores of other ectomycorrhizal fungi are likely to be blown into a truffière, perhaps from trees on a neighbouring property, or spores of another truffle may be deposited in a parcel of dung by a passing pig. Not much can be done to avoid this apart from fencing out wandering animals, including curious humans wearing muddy boots.

Soil Texture and Drainage Characteristics

Because truffles spend almost all their lives surrounded by soil, except when their spores are being dispersed, it is important to understand just a little about soil structure and chemistry. The mineral components of soils are a mixture of very fine clay particles that are less than 0.002 mm in diameter, medium sized silt particles between 0.002 and 0.05 mm in diameter, and sand particles between 0.06 and 2.0 mm in diameter. Some truffle soils may also contain limestone gravel, stones, and rocks that can be very important in buffering the soil pH over decades. A soil's physical properties and its suitability for cultivating truffles are largely determined by the proportion of clay, silt, and sand in a soil; the parent material(s) these are derived from; and accumulated organic matter. The common feature of almost all truffle soils is that they are either naturally well aerated or are maintained in this state by working the ground with machinery. Truffles really don't like having their feet wet nor do they survive well in soils that quickly dry out over the summer months.

According to surveys conducted in France, Italy, and Spain, the Périg-

ord black truffle and Burgundy truffle are found in free-draining, well-aerated soils with good natural drainage throughout the soil profile. The soils do not have an excess of silts, sands, or clays. The Burgundy truffle tolerates soils with a little more clay than the Périgord black truffle, whereas the bianchetto can tolerate very sandy soils.

It is possible to get some idea of how much clay and sand there is in a soil by rolling a small quantity of soil in your hand with a little water—spitting on it is a trick soil scientists use when out in the field and well away from a tap. If there is a large proportion of clay in the soil, it is possible to make a small rubbery strand that will resist breaking. This will not happen if there is relatively little clay in the sample, and instead the soil may feel gritty when rubbed between your fingers. For the inexperienced, however, it is best to have a soil textural analysis performed that measures the proportions of clay, silt, and sand. This information can then be placed on a soil textural triangle that has the normal textural range marked on it for a particular truffle (see Chapter 4). For example, a soil that is 30 per cent clay, 30 per cent sand, and 40 per cent silt is a clay loam, should be suitable for either the Périgord black truffle or the Burgundy truffle.

We are often asked if it is possible to modify a soil if the texture is wrong. Our answer is usually diplomatic and along the lines that anything is possible. However, such a task is anything but trivial. For instance, just adding 10 per cent of sand to the top 20 cm of soil would mean adding about 20 kg/m^2 or the equivalent of more than 10 large truckloads of sand per hectare. Thus, in reality it is best to find a site that has suitable soil characteristics.

In addition to the mineral content of soils, there is also organic matter from rotting plant and animal remains. Typically, Périgord black truffle soils will contain up to 8 per cent organic matter, Burgundy truffle soils somewhat more, and desert truffle soils often much less. Some of this organic matter will have been well decomposed by fungi and bacteria, whereas other components such as newly fallen leaves will not have been incorporated into the soil. Organic components of the soil are of vital importance to the makeup of the soil biota, including the ectomycorrhizal fungi, but our knowledge and understanding of how they interact is sadly lacking and restricted to just a few scientific studies.

Soil Nutrients

Not all of the various nutrients found in a soil are completely available to plants. For example, most of the phosphorus is bound very tightly to soil particles, and even the fine threads of a mycorrhizal fungus cannot access this element. Consequently, to estimate how fertile soils are soil chemists have had to develop ways of extracting and measuring the approximate quantities of elements that are available to plants. Unfortunately, tests done in one part of the world can be very different from those used elsewhere and cannot be easily compared. In addition to phosphorus, other elements that are particularly important to plant growth and in truffle cultivation are calcium, magnesium, nitrogen, potassium, sulphur and, most importantly, the trace elements such as boron, copper, iron, manganese, and zinc, which are required in tiny quantities but can make the difference between a healthy and a very sick plant.

Typical Périgord black truffle (*Tuber melanosporum*) soils are rendzinas and brown earths with high concentrations of plant-available calcium and magnesium, moderate levels of phosphorus, and relatively little sodium. Imbalanced soils in which the ratios of phosphorus to calcium, phosphorus to potassium, and potassium to magnesium deviate from normal for the type of soil are rarely productive for this species. High calcium levels and a high pH can also interfere with the uptake of other nutrients by host plants, particularly the trace elements boron, copper, iron, manganese, and zinc. Because this may lead to chlorosis and poor growth, particularly in the common oak (*Quercus robur*), soils with a pH higher than 8.0 are not recommended. In France it is said that the best truffles come from areas that have iron-rich, red limestone.

Soil testing can be an expensive exercise, so care needs to be exercised to ensure that soil samples are collected properly. Small atypical areas within a site should not be sampled, such as those near water troughs, gates, headlands, trees, dung or urine patches, abnormally wet or dry areas, or other areas with abnormal fertility, such as stock camps. Judicious sampling will also keep costs to a minimum. For the best soil samples, use the following steps.

TOP Yellowing of the foliage (chlorosis) with deeper green midveins, in this case due to iron deficiency, is similar to the damage caused by a lack of boron, copper, manganese, or zinc. I. R. Hall

BOTTOM The generalized foliar yellowing of these hazelnut trees was due to copper deficiency, which was corrected by applying a very small quantity of copper sulphate to the soil. J. Weston

- While avoiding obviously atypical sites, randomly select 5 to 20 spots within the proposed truffière. Five samples is the bare minimum required for a plantation containing 20 plants.
- At each spot remove the vegetation and the litter layer, but no soil, by using a sharp spade held parallel to the soil surface. Take a sample of soil down to 10 cm with a trowel, and place it in a bucket. Repeat this step at each sampling spot.
- Thoroughly mix the 5 to 20 samples in the bucket and place it in soil sample bag. Label this bag "0–10 cm."
- Repeat this process at the 5 to 20 spots, but this time remove soil from a depth of 10 to 20 cm. Place the mixed soil in a separate bag labelled "10–20 cm." Do not let soil from the 0- to 10-cm layer contaminate this second sample.

A soil that has been cultivated in the previous few years will tend to be reasonably uniform in the top 20 cm, so separate samples do not need to be taken at the two depths. It is sometimes helpful if a 20- to 30-cm soil sample is taken as well, particularly where there is a limestone-rich subsoil, as this could affect the advice given by a soil scientist on the preparation of the ground. If the area to be planted is large, say more than a hectare, or if the soil is uneven in appearance, the area should be divided into sections that appear similar and the above sampling procedure carried out for each section.

It is important that the soil be analyzed as soon as possible after sampling, because soil chemical characteristics can change quite markedly, particularly if the soil is warm and wet and kept in a closed plastic bag. Consequently, samples should be sent immediately to a soil-testing laboratory where a truffle advisory service is available. Soil samples sent outside a country are likely to require an import permit to guard against the entry of unwanted pests and diseases. Double bagging may also be required to limit the chance of soil leaking from the bags before the sample gets to a laboratory, where special precautions will be taken to handle the possibly contaminated soil. The interpretation of soil analyses is best entrusted to those with experience both in soil analysis and truffle cultivation. A few of these

organizations in Australia, Europe, New Zealand, and North America are listed in the back of this book.

Site Preparation

High-pH soils can be difficult to cultivate—too wet and the soil can be like porridge, too dry and it can be like trying to cultivate concrete. Consequently, when the weather is either too wet or too dry much time may be wasted in waiting for the soil to reach just the right moisture content for cultivation. The aim in soil preparation is to produce a well-aerated soil with physical and chemical characteristics similar to those where truffles grow naturally. Preparation of the ground needs to begin up to two years before the scheduled planting date, particularly where the pH of the soil needs to be modified (see below). Most truffles require an open-textured, well-aerated soil, so tractors that do not compact the ground are favoured. Those fitted with balloon tires are particularly good because the wide tires spread the weight of the tractor over a larger area.

The first step in cultivation should be the removal of the existing vegetation, whether it is a vineyard, a pasture, scrub, or forest. Removing as much undecomposed organic matter as possible before cultivation has the advantage of minimizing its acidifying effects as it rots in the soil and produces organic acids. If there are ectomycorrhizal trees in the area, removing the trees along with their roots will simply spread ectomycorrhizal contaminated material over an even wider area no matter how carefully the operation is carried out. An excess of dynamite used to remove a willow proved to be a very effective way of spreading *Tuber maculatum* around a New Zealand truffière. While the owners found this exciting, it can hardly be recommended. In a truffière in Texas they even went as far as removing pieces of oak root by hand in an attempt to reduce the level of competing fungi in the soil. While this might give a feeling of confidence that ectomycorrhizal fungi are being restricted, in reality there are likely to be many microscopic spores of competing ectomycorrhizal fungi left behind in the soil.

Ectomycorrhizal trees can be cut off just above ground level and the roots poisoned, or the plants can be left in place and the truffière planted

Balloon tires will distribute the weight of a tractor over a larger area and prevent soil compaction.
Courtesy of PowerFarming

Mowers, like this Grillo Bee Fly, that remove the clippings will reduce the amount of organic matter that rots on the soil surface and would have a tendency to lower the soil pH.
Courtesy of Agrigarden Distributors

Steep land in the Apennines is often terraced, with the truffle host trees planted on the terraces and shrubs planted on the banks to limit soil erosion.
D. Grammatico

around them, but far enough away to minimize any risk of contamination, say 50 m. For the same reason, truffières need to be established far enough away from nearby forests that might contain, for example, winter truffle. Scrub can be removed with bulldozers, a grader, or a tractor fitted with a rear-mounted blade, but taking care not to remove the valuable topsoil. Because grape vines, fruit trees, and the like are arbuscular mycorrhizal plants, they can be ripped out and carted away without any problems from competing ectomycorrhizal fungi. Pasture can be grazed off or mown and the clippings removed, and the remaining vegetation can be sprayed out with a suitable herbicide such as glyphosate. Large rocks that are likely to damage farm machinery may need to be removed, but there are some serious machines available that are capable of pulverising even quite large lime-

Rudimentary terracing produced with a walk-behind rotary hoe in the GGC truffière, North Canterbury, New Zealand. The natural tussock vegetation and introduced grasses maintain the integrity of the banks between the terraces. I. R. Hall

stone rocks. This was done in the Arotz truffière in northeastern Spain after the native juniper and holm oak vegetation was removed.

Mobile, free-draining soils with relatively high rainfall, such as on steep slopes in the Italian Apennines, are ideal conditions for truffle cultivation. In these areas terracing, contouring, or making provisions to control erosion are essential components of truffle cultivation. Machinery can be used to terrace slopes of more than 15°, and small rotary hoes can achieve a good result if carefully run across a hill. At the same time it may be possible to incorporate or expose some high-pH subsoil and bypass the need to apply lime to raise the soil pH, although it is important not to lose too much topsoil during this process. Because of erosion problems with soils on steep slopes, dry stone walls can be used to terrace slopes, or the banks between the flattened areas may be stabilized by allowing vegetation to grow on them. Where the land has not been contoured prior to planting, branches and rocks are sometimes placed along the contour of the slope to prevent soil being lost and the host trees' roots being exposed during heavy rainfall.

Modifying soils

Lime is applied to soils in France in an attempt to raise the soil pH to a point where it will favour growth of the Périgord black truffle (*Tuber melanosporum*) and disadvantage that of the winter truffle (*T. brumale*). In New Zealand and other parts of the Southern Hemisphere, high-pH soils are not as common as they are in Europe. Thus, techniques had to be developed for modifying soils to make them suited to the cultivation of the Périgord black truffle. For soils with a relatively low pH that overlay high-pH subsoils or shattered limestone, it was possible to raise the pH simply by deep ploughing followed by cultivation with tines. For low-pH soils that overlay low-pH subsoil, however, the solution was the incorporation of very large quantities of fine agricultural lime (calcium carbonate).

The quantity of lime applied depends on the natural pH of the soil and its buffering capacity—the intrinsic ability of the soil to resist changes in pH. Typically, the amount of fine limestone applied to soils in New Zealand has been 1.0 to 1.5 tonnes/ha for each 10 cm of soil treated and for each

0.1 unit that the pH had to be raised. For example, to raise a soil to a depth of 30 cm from pH 5.9 to 7.9 requires 60 to 90 tonnes/ha. Additional hard limestone chips between 5- and 20-mm diameter can also be mixed into the soil to help buffer the pH once it has risen to the ideal. Even so, additional lime may need to be added to the soil as the calcium migrates down through the profile with passing years. Occasionally, faster-acting, but more expensive and caustic hydrated lime or quicklime was applied by those in a hurry, while others have used a mixture of dolomite and lime in areas where the available magnesium concentration had been low to begin with.

Once the pH of a soil has remained stable for two years, it appears that maintenance applications of lime will be required once or twice a decade rather than annually. Generally, the adjustment of pH has taken from 6 to 18 months, but it may take two years or more. As the pH of soils rises, they become more friable and freer draining as the clay particles flocculate. However, a high pH is often accompanied by expected falls in plant-available concentrations of the trace elements boron, copper, iron, manganese, and/or zinc. Occasionally, these may drop below threshold concentrations and plants show deficiency symptoms of yellowing leaves. For this reason, applying excess lime (or bringing too much up from the subsoil) should be avoided by regularly testing the pH and making several judicious applications of lime so as to raise the pH in steps.

The most ambitious modification of a soil in New Zealand has been that of an acidic soil overlying volcanic ash on Don and Isabel Dempsey's truffière south of Taumarunui, in the North Island, where the pH was raised from 5.3 to 7.6. This and six other New Zealand truffières on naturally acidic soils have produced truffles.

Rotary hoes are very useful machines for initially working lime into a soil, but they must be used judiciously. If used more than a few times, rotary hoes can destroy the soil structure and turn it into something with the consistency of talcum powder. They may also polish the soil at the lowest point they reach, which can impede drainage. A mechanical spader pulled by a powerful tractor is also an excellent tool and in one or two passes can incorporate most of the lime that is needed to modify the pH to a depth of 30 cm or more.

TOP In New Zealand, soils that have developed over recent volcanic ash are very free draining and, after liming, can produce large quantities of truffles. I. R. Hall

BOTTOM This mechanical spader can work large quantities of lime into a soil to a depth of 30 cm or more with just a few passes. I. R. Hall, courtesy of P. Harris

Fertilizers

Soil tests may identify deficiencies in nutrients, such as potassium, sulphur, and trace elements. In general, however, the only element worth applying prior to planting if it is in short supply is magnesium, which can be applied as dolomite. Other elements can be applied after planting, when it is possible to confirm any likely deficiencies after performing nutrient analyses of the foliage. These foliar analyses are more accurate than soil analyses.

Fencing

Stock, rabbits, hares, pigs, and opossums can create problems in a truffière, particularly during the establishment phase. Equipping fences with electric outriggers placed on the outside of the fence 750 mm above the ground and 150 mm out from the fence can be effective in controlling opossums and goats. In Europe, wild boars rooting for truffles can be a major problem in productive truffières, and consequently stout fences are recommended to keep them out. In the Arotz truffière in northeastern Spain, the fence is actually concreted into the ground to prevent pigs digging under it. Rabbits are more easily discouraged. Running rabbit netting up the outside of a fence and out 30 cm over the surface of the soil from the bottom of the fence should be sufficient to stop them getting in.

Tree guards and windbreaks

Protection from the desiccating effects and physical damage caused by wind is known to have beneficial effects on the growth of many tree crops. Various tree protectors are widely used in the agricultural and horticultural industries. In addition to providing protection from wind and herbicide sprays, they also offer some protection from browsing animals, such as rabbits and hares. The 15-cm square by 60-cm high KBC tree shelters are wide enough to allow good air circulation inside the boxes, thus minimizing the possibility of the plants being cooked during hot, dry weather. Also, they do not force the trees to grow up too fast and become spindly, a problem we have encountered with oaks inside 90-cm high by 10-cm diameter tubes. Alternatively, there are an array of tree protectors on the market made from fine to coarse mesh that also allow for good air circulation.

TOP A KBC tree shelter (15 cm by 15 cm by 60 cm). The straw mulch, used to conserve soil moisture, has been covered with wire mesh to stop it from blowing away in the wind. I. R. Hall

BOTTOM The Tubex shrub shelter is an alternative to the KBC tree shelter. Tubex

In Haute Provence truffières are not generally established on slopes exposed to the cold northerly winds. Similarly, in southern Provence truffières are not generally established on south-facing slopes exposed to dry southerly winds. Permanent windbreaks do not feature around many European truffières primarily for economic reasons but also because these are less important in Europe than in parts of New Zealand, where the wind run is considerably higher (wind run is the distance that wind travels over a particular period of time). Windbreaks can also raise air temperatures by about 0.6°c above ambient within a zone on the leeward side about six times the height of the windbreak. They can have an even greater effect on soil temperature, and in areas that are climatically marginal for truffles, windbreaks may make the difference between success and failure.

Living shelter is much cheaper than artificial shelter, but considerable forward planning is needed to ensure that it is tall enough to have an effect when it is needed most during the early years of a truffière. It is also important not to use plants that can harbour competing fungi. Appendix 1 provides a list of some plants that form arbuscular mycorrhizas and do not present problems, and Appendix 2 lists plants that form ectomycorrhizas and should not be planted adjacent to a truffière. Unfortunately, many of the most useful and fast-growing trees that are normally used in shelterbelts, such as poplars and pines, are ectomycorrhizal and therefore unsuitable. Because many plants struggle to grow in high-pH soils, if a truffière is to be established on a naturally acidic soil, not applying lime in those areas destined to be planted with a windbreak will allow a wider choice of trees.

Although artificial windbreaks (4 to 6 m high) have been used around some New Zealand truffières, they were expensive to erect and maintain. Care also had to be taken to ensure that they would perform as required and positioned so that they did not shade the root zones of the truffle trees during the growing season. This is particularly important during spring and in cooler locations where the soil temperatures must be maximized. A few windbreaks were removed once the oaks reached 5 m high, after about five years. By this stage the trees were sufficiently robust to withstand the effects of strong winds that are characteristic of some areas.

TOP Cypresses, which form arbuscular mycorrhizas, are commonly used in New Zealand to produce effective windbreaks. I. R. Hall

BOTTOM The barnea olive (*Olea europaea*) grows quickly in warm locations and makes an effective and productive windbreak. I. R. Hall

A 4.5-m artificial windbreak surrounded the first New Zealand truffière to produce truffles. The willow tree outside the truffière on the back boundary is thought to have been the source of the *Tuber maculatum* that competed with the Périgord black truffle before it began fruiting. I. R. Hall

Planting Density and Tree Arrangement

The Périgord black truffle (*Tuber melanosporum*) can thrive and fruit in a wide range of climatic conditions, from cool temperate to Mediterranean and even subtropical (Appendix 7). This species' ability to tolerate extreme conditions and to continue to dominate even under a partial canopy has given it access to habitats that other fungi would not be able to tolerate. But it would be foolish to think that a single planting density and arrangement would suit the Périgord black truffle, let alone all other species of truffle. For example, the Italian white truffle (*T. magnatum*), which is adapted to a relatively moist soil and equitable conditions, would be unlikely to survive, let alone fruit, in the rigorous conditions provided by a bare soil under a blistering Mediterranean sun and with limited irrigation. Similarly, because soil moisture, temperature, fertility, and drainage characteristics are likely to change as a truffière ages, a single planting density cannot be expected to suit both young and old truffières. Consequently, some care

and thought are needed when planning the layout of a truffière, particularly in those areas where there is no local knowledge to fall back on.

Soil temperature and moisture are affected by the ambient temperature, wind run, and how much sun strikes the ground. Lower planting densities might therefore be most appropriate in cooler areas, particularly where there is irrigation and a fertile soil that encourages growth. However, if the planting density is only 100 trees per hectare, a planting density recommended by some for the Périgord black truffle in France, valuable land will be wasted until the tree roots and their fungal partner fully colonize the soil—perhaps one or two decades after planting. In Alain Monnier's successful truffière in Marigny-Marmande near Richelieu (47°N, less than 200 m elevation), which is at the northern tip of Périgord black truffle production in France, the trees are planted in widely spaced rows running more or less north–south and most of the soil between the rows is heated by the sun for most of the day. Similarly, in the 600-ha Arotz truffière (Soria, northeastern Spain, 42°N, 1250 m elevation), with 280 trees per hectare and a 6 m by 6 m spacing, there is relatively little shading.

At the other extreme, a planting density of 800 trees per hectare was recommended by Jean Grente in the late 1970s for the Périgord black truffle. However, with this number of trees per hectare, the rows 4.5 m apart running east–west, the trees 2.8 m apart within the rows, and at a latitude of 45°N, very little of the ground would be heated by the sun even during midsummer once the trees grew to 3.5 m high. This arrangement was used in Alan and Lynley Halls' truffière, Gisborne, New Zealand (38°s), and initially it proved very successful. After canopy closure, however, competing fungi began to depress production and yields only recovered after half the trees were removed.

TOP A low planting density was used in Alain Monnier's truffière, Marigny-Marmande, France, with alternating rows of holm oak (*Quercus ilex*) and pubescent oak (*Quercus pubescens*). I. R. Hall

BOTTOM Relatively little shading occurs at any time of the year in the 600-ha Arotz truffière near Soria, Spain. During the preparation of the ground, heavy machinery was used to break up large limestone rocks. I. R. Hall

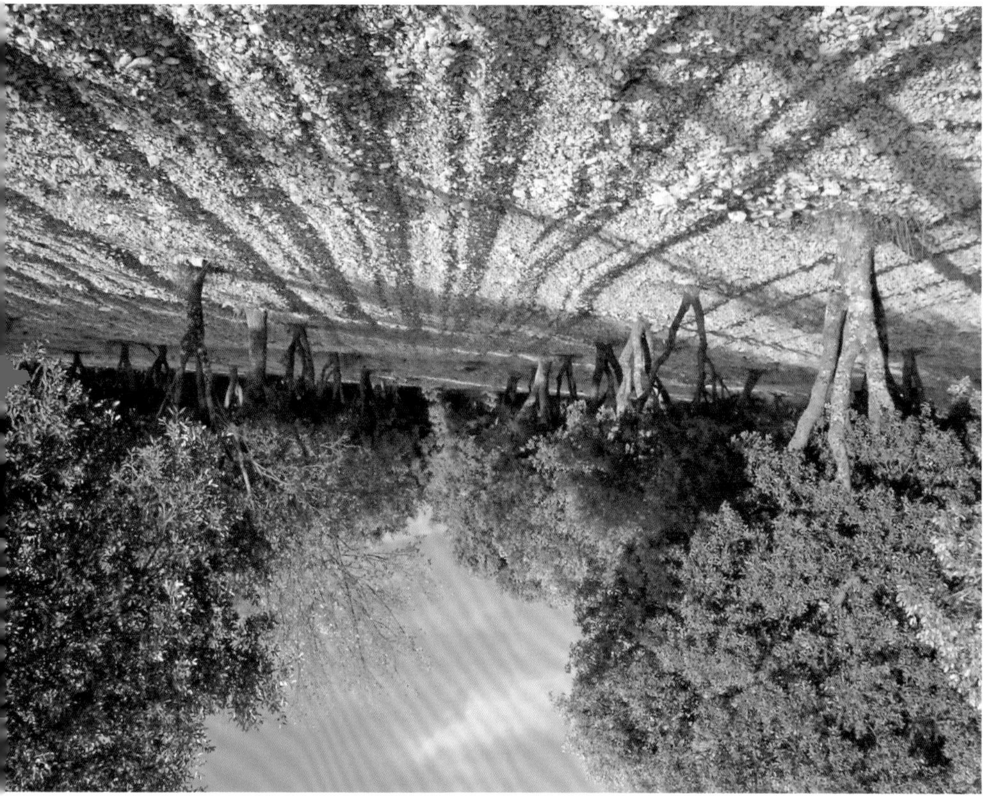

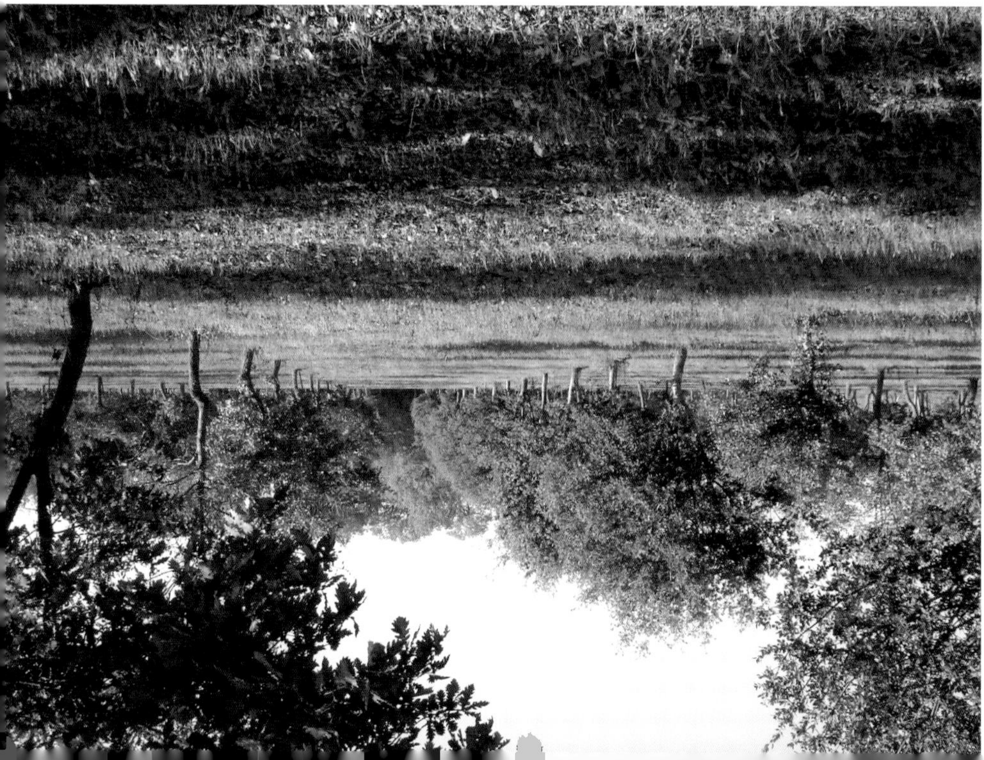

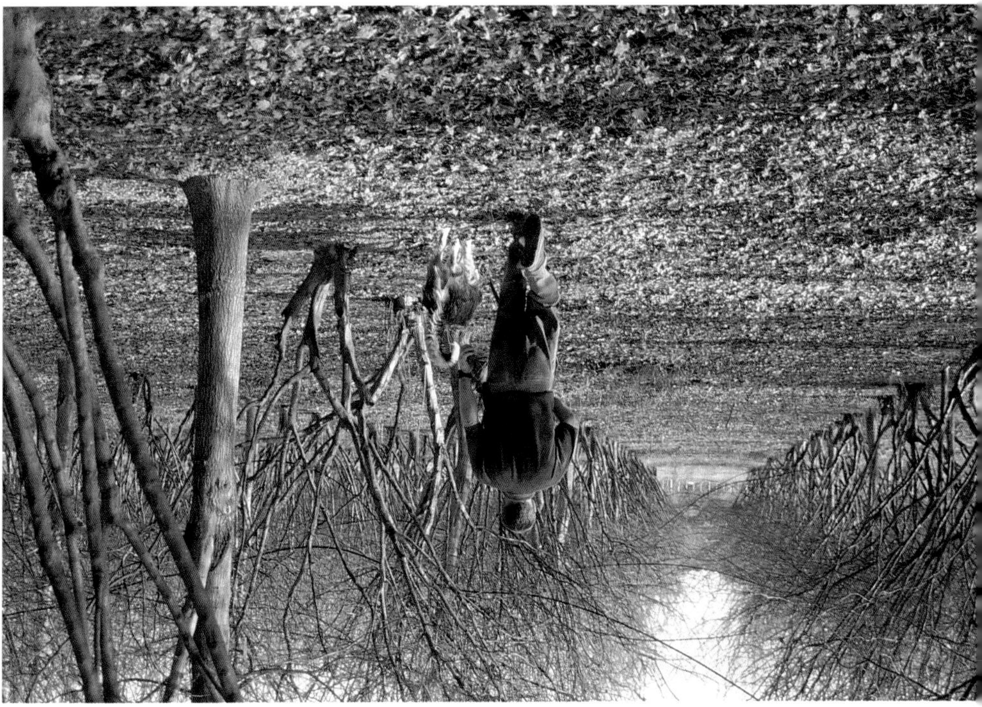

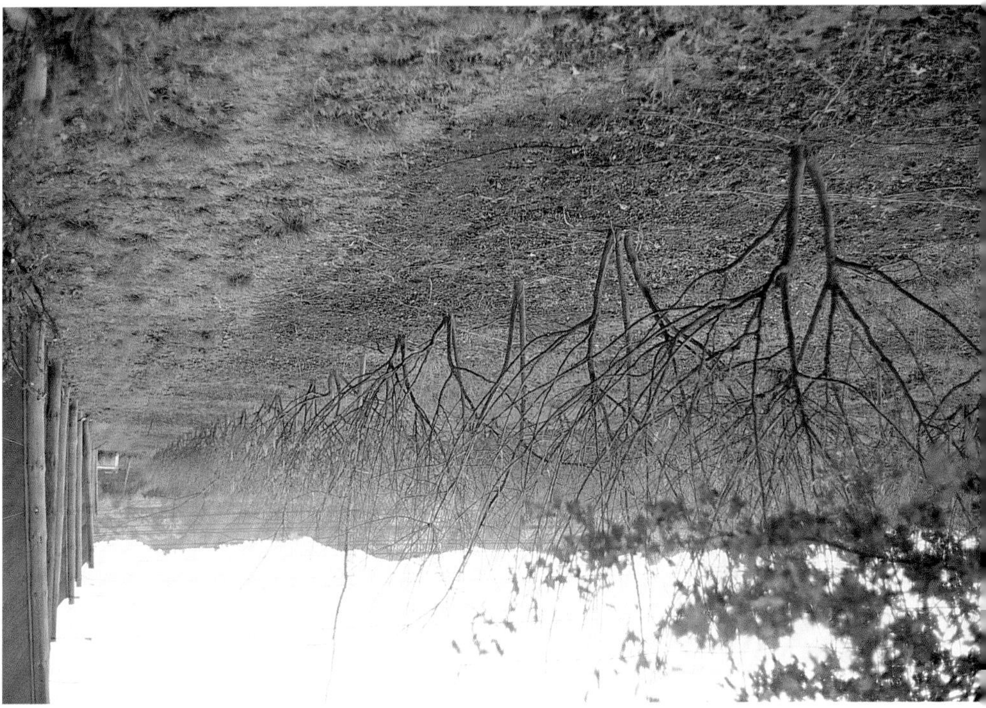

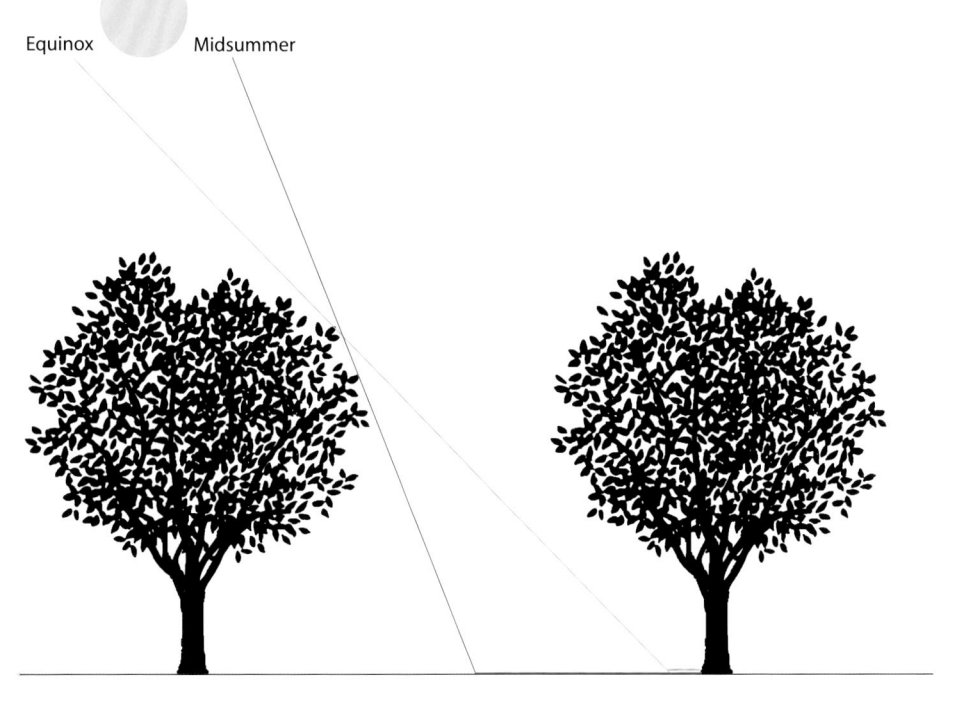

Equinox   Midsummer

TOP LEFT Alan and Lynley Halls' Périgord black truffle (*Tuber melanosporum*) truffière when five years old and a few truffles had formed. At this time the brûlés had almost covered the truffière. The planting density was 800 trees per hectare. I. R. Hall

BOTTOM LEFT Alan and Lynley Halls' Périgord black truffle (*Tuber melanosporum*) truffière when 11 years old and 2 years after the start of commercial production. A year later production reached more than 100 kg/ha but then declined when the canopy became very dense. Plant density was then reduced to 400 trees per hectare. I. R. Hall

ABOVE At a latitude of 45°N and during the middle of the day, little of the ground would be heated by the sun during midsummer in truffières with trees 3.5 m high, the rows 4.5 m apart, and the rows running east–west and planted close enough to almost form hedges. At the equinox, almost none of the ground would see the sun even during the middle of the day. I. R. Hall

A single species of host plant arranged on a grid is the simplest design. There may be some justification for arranging such a truffière with a hexagonal, close-packing arrangement (trees on each corner of a regular hexagon with one plant in the centre) to increase the efficient use of the land by 15 per cent over a square grid. However, it is probably much more important to ensure that the plant density does not exceed the availability of water at the height of summer.

Many other designs have been suggested for Périgord black truffières, such as alternating widely spaced rows of holm oak (*Quercus ilex*) and pubescent oak (*Q. pubescens*), alternating pairs of hazelnuts (*Corylus*) with one oak, and alternating the positions of oaks between the rows, as well as quite complex designs where four or more host plants are carefully spaced out within the truffière. In parts of the world where there is no tradition of truffle cultivation, growers have no yardstick and have difficulty deciding what a suitable planting density and arrangement might be. Consequently, some in New Zealand have experimented with designs with uneven spacing giving planting densities between 400 and 800 trees per hectare. There are many other possibilities when it comes to locating a truffière on a property, such as planting either side of a driveway or around the perimeter of a field between double fences.

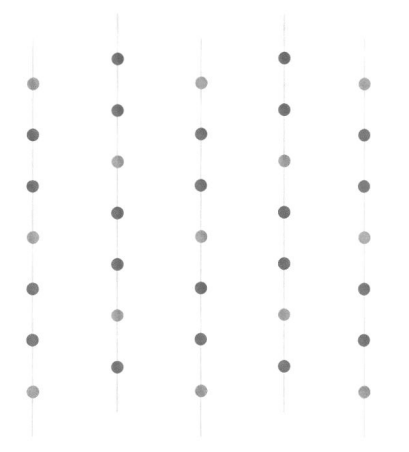

A successful planting arrangement widely used in New Zealand has been to alternate each oak (red dots) with two hazelnuts (blue dots) within rows and stagger the position of the oaks between the rows. For 400 trees per hectare a suitable spacing is 6.5 m between the rows and 3.8 m within the rows, and for 800 trees per hectare 4.4 m between the rows and 2.9 m within the rows. I. R. Hall

A variable planting density can be achieved by planting blocks of trees at 800 trees per hectare and separating the blocks by double-spaced rows.
I. R. Hall

High planting densities are used in the cultivation of the Burgundy truffle (*Tuber aestivum*). This one is the second Burgundy truffle truffière to produce in France. It was planted in autumn 1976, at Mont-Dore, Haute-Saône, Franche-Comté. By 2001 the trees had formed an archway. G. Chevalier

We believe that planting densities for the bianchetto truffle (*Tuber borchii*) should be similar to those used for the Périgord black truffle. Such advice should be treated with some caution, however, because the cultivation of bianchetto is still very much in its infancy. Because the Burgundy truffle (*T. aestivum*) grows well under nearly closed canopies in Gotland, Sweden, and in France, higher planting densities (up to 1250 trees per hectare) are used when compared with the Périgord black truffle. Even higher planting densities are used in the cultivation of the desert truffles in southern Spain, where around 1400 rockrose (*Helianthemum almeriense*) per hectare are planted. Although the Italian white truffle (*T. magnatum*) has yet to be cultivated with any degree of success, Italian researchers suggest planting at a relatively high density, similar to that used for Burgundy truffle, to ensure an early canopy closure and then removing excess trees later.

Irrigation System

In southern Europe warm springs and hot dry summers are punctuated by brief, heavy rainfalls. In summer this rain appears to be part of the trigger for truffle initiation. Adequate summer and autumn rain ensures high yields, while in years with little rainfall truffle yields are severely depressed. This information is embodied in French and Italian folklore (water in July, truffles at Christmas), and the experienced truffle hunter of 200 years ago, as now, took great care to ensure that soil moisture was conserved during periods of drought. Mulching with straw or placing juniper branches over the brûlés were popular practices. Black polythene and other mulches have also been used in Europe but with mixed results (Chapter 6).

Despite the knowledge that adequate soil moisture is essential for truffle formation, irrigation systems are not common in France, Italy, and Spain. This is partly because irrigation water is either not available or is a scarce and expensive commodity. Truffle growers can also be somewhat conservative, and an irrigation system is still considered avant-garde by many. In all the areas in New Zealand where truffles may be grown, there is the possibility of periodic drought. Consequently, some form of irrigation equipment has been installed in almost all of the country's 150 truffières. These

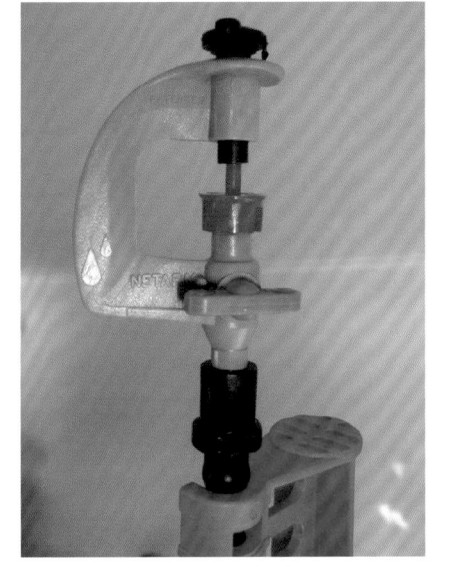

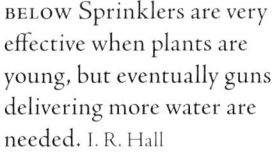

LEFT A sprinkler with interchangeable irrigation heads is suitable for irrigating individual trees. I. R. Hall

BELOW Sprinklers are very effective when plants are young, but eventually guns delivering more water are needed. I. R. Hall

irrigation systems range from domestic hose-pipes for small truffières to pumped or gravity-fed schemes with computer control and sophisticated filtering systems to remove excess manganese or iron that might otherwise block the sprinklers.

Obviously, there is little point in irrigating much more than the area of ground occupied by the host plant's root system, as this would simply

encourage the growth of weeds that would then have to be controlled. In the first year only rarely would the root zone extend beyond the edge of a circle 1 m in diameter, and consequently applying 20 mm of irrigation water to 500 plants would require no more than 8000 L/ha of water per irrigation. Of course, as the plants grow so too will their root zones, and the total amount of water required for irrigation during a serious drought would increase to in excess of 100,000 L/ha per irrigation. On some truffières in Europe and Australasia, the problem of irrigating an increasing root zone has been tackled by installing microjets or mini-sprinklers with interchangeable heads. Initially, heads are installed that deliver only low volumes over a relatively small area, but as the plants age these are replaced with heads that irrigate larger areas. Other irrigation systems have heads that sprinkle water down onto the soil over a relatively small area but can be easily inverted so that the water is forced up and outwards to irrigate a larger area.

The use of drippers is not recommended because it can cause bunching of the mycorrhizas, which is to be avoided. However, if drippers are used, several must be installed per plant and arranged so that they distribute the irrigation water evenly over the rooting zone. Once the trees grow to more than a few metres high and the roots zones of the trees start to touch each other, much larger irrigation heads can be installed. Those that might damage the foliage with the force of water are best avoided, as this might allow the entry of disease organisms into the plants.

Black polythene irrigation tubing heats up in the sunshine to the point where the water inside can approach the boiling point. Clearly this would not benefit the health of plants watered with it. One way to avoid this problem is to completely drain the irrigation lines after each irrigation—a simple task with truffières established on a hillside but not at all easy to achieve on flat land. Another way is to bury the main irrigation lines using, for example, a trencher or a mole plough. However, where the pH of the topsoil has been raised by the application of lime, the trencher will lift relatively low pH subsoil to the surface that would then have to be limed again after the pipes had been laid. Other ways include using a system of valves that allows hot water in the pipes to be flushed out before irrigating or simply irrigating at night after the water in the irrigation lines has cooled

A trenching machine used for burying pipes can also be used for cutting slots through the ground to stop roots from neighbouring trees entering a truffière. I. R. Hall, courtesy of Hirepool

A solenoid irrigation valve operated by a 9-volt battery can be useful where mains power is not available. These valves can also be useful where there is a lack of irrigation water to irrigate all the truffière simultaneously or when hot water in the irrigation lines needs to be flushed out before irrigating. I. R. Hall

down. Such a system could be easily automated by using battery-operated solenoid valves. The installation of automated dosing systems would be useful on large truffières, where trace element applications are likely to be needed. Similarly, filters would be a useful addition where there is particulate matter in the irrigation water, whereas special water treatment systems including aeration, ozone treatment, or special filtration may be required

to remove high concentrations of iron and manganese that can precipitate inside and obstruct irrigation jets.

Competing Ectomycorrhizal Fungi on Nursery Stock

Pulvinula constellatio and *Sphaerosporella brunnea*, two species of Ascomycetes, are commonly found contaminating truffle-inoculated plants in nurseries. Through the production of masses of tiny air-borne spores that form in small cup-shaped fruiting bodies on the surface of the potting mix, these fungi can rapidly colonize all the plants in a greenhouse and prevent infection by the inoculant truffle fungus. Fortunately, these fungi usually die out once the seedlings are transplanted to the field, and if the seedlings still have truffle infections on them, the truffle fungus can recolonize newly uninfected root. Basidiomycetes can also present significant problems for the producer of truffle-infected plants. The most common of these are species of *Thelephora* and *Scleroderma*. These present special problems because not only are they very infective, but it is possible for the novice to confuse their infections with those produced by a truffle.

In Europe, the poor-flavoured winter truffle (*Tuber brumale*) is one of the most competitive ectomycorrhizal fungi in truffle cultivation, and its spread through the truffle-producing regions has been regarded as a catastrophe. Part of this problem can be laid at the feet of scientists and nurseries. In the 1980s and 1990s, Périgord black truffle growers in France were urged to plant infected hazelnuts instead of oaks because they grew faster and produced earlier. Unfortunately, the scientists hadn't done sufficient research. Jacques Pébeyre, the senior owner of one of the leading truffle trading companies in France, was quoted in the book *From Here You Can't See Paris* (Sanders 2003): "Les scientifiques, they can do good, but for a long time they would have us plant hazelnut trees instead of truffle oaks. And now all those truffle grounds are infested with the low quality brumale variety and not the black truffle. Why did the scientists tell us to plant hazel trees? Because they grow faster! With a truffle oak, it takes ten years until they are ten feet high. With a hazelnut, two years. The scientists didn't think it mattered what kind of tree you planted [but] they were wrong!"

LEFT *Sphaerosporella brunnea* fruiting bodies are a common contaminant in greenhouses during the production of truffle-infected plants. A. Zambonelli

RIGHT *Thelephora* can be a major ectomycorrhizal contaminant in greenhouses where truffle-infected plants are being produced. I. R. Hall

Part of the problem was that the inoculum that nurseries used was contaminated with the highly competitive winter truffle. This fungus proliferated on hazelnuts at the expense of the preferred Périgord black truffle and, after outplanting, contaminated large areas of French truffières. In addition, many truffières were established adjacent to pubescent oak (*Quercus pubescens*) forests, a good host for the winter truffle, which spread into the truffières by animals, on the roots of trees spreading out from the forest, or simply on the boots of people walking through the forest and into the truffière.

A decade ago the only major contaminating truffle was the winter truffle, but in recent years the European market has been inundated with very cheap Asiatic truffles that sell for just a fraction of the price paid for the Périgord black truffle. These species are morphologically very similar to Périgord black truffles and have fooled truffle-infected plant producers, as well as chefs and gourmets, into thinking they were buying truffles of European origin. Similarly, in Italy plants sold as inoculated with Italian

white truffle are commonly contaminated with bianchetto, *Tuber maculatum*, and/or *T. dryophilum* probably due to the accidental inclusion in inocula of incorrectly identified rotting truffles of these highly infective species. It seems a joke of nature that the most competitive species are those that are least appreciated as foods.

The threat posed by the inclusion of the poor-flavoured Asiatic truffles has been taken very seriously by the fledgling New Zealand truffle industry. In an attempt to prevent the entry of these unwanted fungi, the New Zealand Ministry of Agriculture and Forestry recently introduced regulations that require all imported truffles to be labelled and inspected by fungal taxonomists before being released from containment. At the time of writing this book, such regulations had not been introduced by other countries in the Southern Hemisphere, and avoidable contaminants have appeared in some of their truffières.

Perhaps in an attempt to avoid legal problems, or even to hoodwink a client, some producers of plants will hide behind terminology by stating that their plants are "inoculated with *Tuber melanosporum*," that is, the correct truffle fungus has been placed on the plant. There is an important difference between *inoculating* a seedling with a particular fungus, however, and ensuring that the seedling actually becomes adequately *infected* with the correct species of truffle and is uncontaminated with other ectomycorrhizal fungi.

In an attempt to guard against the use of contaminated, poorly infected plants or plants that are simply not infected with truffle at all, several researchers have devised methods of assessing the levels of infection on plants. After being inspected by independent, qualified personnel and found to have adequate root tips mycorrhized by the inoculant truffle and little or no contamination, a certificate is issued. It is only ever possible to inspect a sample of plants, however, and it is very difficult to distinguish some infections, such as those of the Italian white truffle, *Tuber maculatum*, and bianchetto, so mistakes are almost inevitable. Also, the certificate can only ever refer to a batch of plants of the same age, produced with a uniform batch of inoculum, using the same propagation method, and inoculated using a standard technique in the same place. Thus, problems can

arise when a nurseryman, perhaps conveniently, assumes that the certificate covers many thousands of plants in batches the inspector never saw. In Tuscany, Italy, the law actually allows for up to 15 per cent of the root tips of an inoculated tree to be infected by other species. This is partly because of the difficulties of producing uncontaminated plants, but also because ectomycorrhizal fungi are likely to be present in most sites where the trees are planted. However, it is not always justifiable to assume that all contaminants will disappear after outplanting because they will not be able to compete in the field against the truffle. While molecular techniques can be used to screen plants, cost currently prohibits their extensive use.

Regrettably, poor seedling quality remains a serious problem throughout the truffle industry. Large numbers of inoculated plants on the market are either heavily contaminated with other species of truffles, other ectomycorrhizal fungi, or, worse still, are completely free of truffle mycorrhizas. This is of no benefit to the truffle industry and suggests that the suite of techniques used by commercial nurseries is sometimes inadequate for producing plants with good truffle infections and without contaminants. In addition, when the recipes are followed religiously the production costs may make them uncompetitive in the marketplace. A section of this industry may be ignorant of what they are doing, or perhaps they simply don't care. Clearly, would-be truffle growers and investors need to be aware of the risks in trying to economize by finding what appear to be the cheapest source of plants, as well as the long-term benefits of paying to have a batch of truffles destined for their truffière thoroughly checked by an independent specialist to ensure that they are free of contamination. They also need to be aware of promises by fly-by-night operators who "guarantee" their plants only to disappear before the harvest is due.

Other Plant Specifications

Growers also need to ensure that the plants that they obtain from a nursery are free from minor or major plant diseases, an obvious requirement but surprisingly one that can be overlooked by the nurseryman. Unlike island countries such as Australia and New Zealand, where there are strict

import regulations, in the European Union there are very few restrictions on what can be moved around. Recently, however, a system of "plant pass-porting" has been introduced that attempts to limit the movement of the most serious pests and diseases. Some examples of pathogens that are particularly important to the truffle grower are Eastern filbert blight (*Anisogramma anomala*) and bacterial blight (*Xanthomonas arboricola* pv. *corylina*), which affect hazelnuts, and *Phytophthora ramorum* and *Phytophthora kernoviae*, which can kill various species of oak. We describe the control of some potential pathogens in Chapter 6.

Where truffle-infected plants are to be used in reforestation projects, there may also be minimum specifications that have to be met. These may include plant height and collar diameter, although these specifications are less important to those truffle growers who protect their plants from browsing animals, such as rabbits and hares, by surrounding them with protective tubes after planting.

Choice of Host Plant

Appendix 3 lists the host plants for each of the main truffles of commerce and well-recognized hosts for the various species of truffle, as well as other species that have been used in laboratory studies. The most important plant families are Fagaceae, Pinaceae, and Salicaceae, although Cistaceae, Mimosaceae, Betulaceae, Corylaceae, Rosaceae, and Tiliaceae also contain some important host plants. In Europe the most common hosts are the common oak, holly oak, Palestine oak, pubescent oak, Turkey oak, Austrian pine, Turkish pines, hazelnuts, hop hornbeam, European hornbeam, and large-leaved lime. Other experimental hosts, such as Monterey pine (*Pinus radiata*) and sweet chestnut (*Castanea sativa*), which have been infected with truffle under experimental conditions, are not included in Appendix 3 because these showed poor growth and/or symptoms of severe trace element deficiency when planted in high-pH soil.

The common hazelnut (*Corylus avellana*) is the most frequently used host plant in truffle cultivation. Its popularity is due to the tree's vigorous growth, tendency to produce a well-developed root system, and reputa-

tion as an excellent carrier of mycorrhizas and producer of truffles several years earlier than other plants. In New Zealand it was initially believed that hazelnuts would not produce truffles beyond 20 years or so, based on European information and the false assumption that they would cease to grow well. However, many of the hazelnuts in England are hundreds of years old and the remnants of ancient coppiced forests once used for producing, for example, thatching spars, charcoal for gunpowder, and layered hedges. Consequently, it is hoped that if hazelnuts are carefully managed they will continue to produce truffles for much longer than 20 years.

At least two other hazelnuts, *Corylus colurna* and *Corylus heterophylla*, have been successfully inoculated with the Périgord black truffle. In at least one study, more truffles were harvested beneath *C. colurna* and several oaks than beneath the common hazelnut. In 1996 a large trial was established in France to investigate whether variations within and between host plant species had any effect on yields. Initial findings suggested that infections were better on clonal material, but by the end of the 2003 season this experiment had not shown any obvious differences between clones except that oak appeared a better host than hazelnut.

A factor that needs to be considered before including a host in a truffière is the final shape of the tree. A spreading habit, like that of hazelnut, is likely to require more frequent pruning, although contact herbicides can be used to keep hazelnut suckers in check. It is because of this and problems with *Tuber brumale* that influential growers such as Alain Monnier in Marigny-Marmande, who has played a major part in reintroducing the Périgord black truffle to the Loire, no longer includes hazelnut in his truffières and primarily uses pubescent and holm oaks. The common oak in New Zealand can grow by as much as 2 m in the second year after planting and can create shading problems within a truffière, although it has been remarkably productive. Species such as holm oak do not have a rapid growth rate but they produce an extensive root system. Adverse weather conditions such as cold winter snaps may also restrict the use of frost-sensitive plants from areas where there are unseasonable frosts or extreme winter temperatures.

There is considerable justification for using two or more host plants in

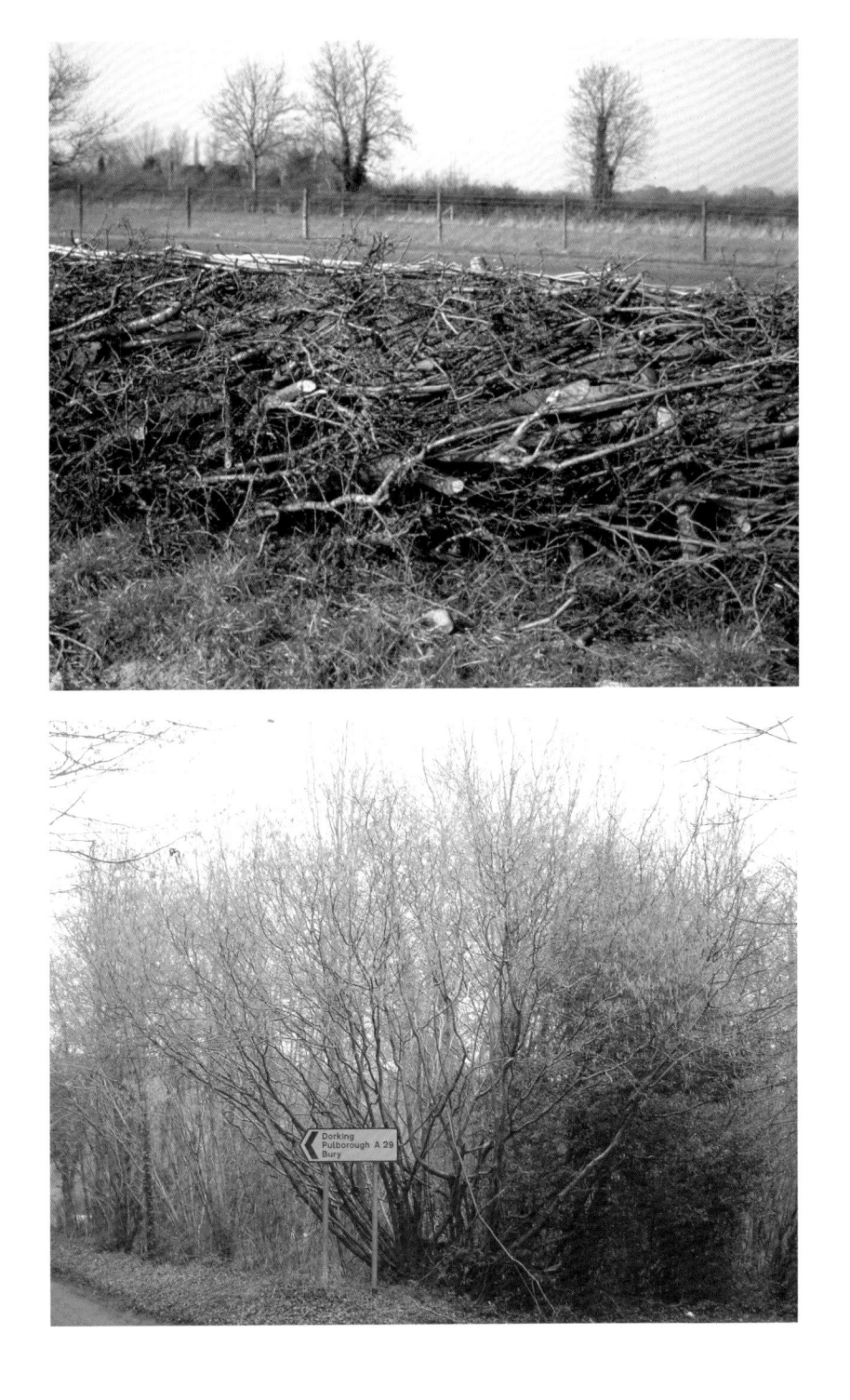

a truffière. In countries where there is no tradition of truffles cultivation, there is insufficient information to decide which of the possible hosts will perform best in a particular location. Of greater importance, perhaps, is that if a pathogen enters a truffière it is not likely to affect all of the species. In the United Kingdom and United States, for example, *Phytophthora ramorum*, the cause of sudden oak death, affects California black oak, cherry bark oak, holm oak, sessile oak, shreve oak, Turkey oak, and European beech. The holm oak, which is a mainstay of the truffle industry in Europe, is particularly susceptible, whereas the common oak is relatively resistant. Unfortunately, both the holm oak and common oak are susceptible to *Phytophthora kernoviae*, which has been found recently in Europe and elsewhere. In contrast, in a 60,000-tree truffière northwest of Austin, Texas, primarily due to fears that oak wilt disease (*Ceratocystis fagacearum*) might cause major problems, only hazelnuts were planted.

For aesthetic reasons, it is better to plant local trees in an area to maintain the natural landscape and ensure that the plants are adapted to the local climatic and soil conditions. Examples include the planting of pubescent oak in the Apennines of Italy, common oak in the deep fresh soil on the planes of northern of Italy, and holm oak in Mediterranean areas of Europe. Similarly, it might be best to inoculate with cultures derived from or truffles collected from the same area they were destined for, just in case there are ecotypes adapted to specific areas. In New Zealand and Sweden, however, Périgord black, Burgundy, and bianchetto truffles have been produced in areas much cooler than where the inoculum originated.

No matter how good a host tree species may appear to be in the field, nurseries will not adopt it if they are unable to produce large numbers of well-infected seedlings, without contamination, and at a competitive price.

TOP The ancient art of layering hazelnut and hawthorn in the United Kingdom has produced hedges that are impenetrable to stock as well as small boys scrumping apples. Some of the hazelnuts and hawthorn in these hedges can be many centuries old. I. R. Hall, courtesy of G. Stephens

BOTTOM These hazelnuts were mature coppiced trees when Ian Hall was still in short trousers 50 years ago. I. R. Hall

Problems that might be encountered are difficulties in establishing infections, perhaps because of insufficient second-order lateral roots that are needed by the fungus to establish mycorrhizas; a depression in growth rate and general vigour around the time the fungus is establishing mycorrhizas; or nonuniformity of infections between plants. The last is very important in the greenhouse because plant uniformity should help minimize the need to screen out poorly infected plants prior to sale—a very expensive task. Also, nonuniform plants will look like a dog's breakfast after outplanting and, more importantly, may behave variably in the field such as when exposed to competing fungi.

Double Cropping and Interplanting

In a truffière where the rows might be as much as 10 m apart, there is a considerable proportion of the land that would be completely unoccupied by the root systems for many years. On dry, marginal lands such as those used to cultivate desert truffles, there appear to be few viable options for double cropping. Lavender is a possibility, as are other species that tolerate a high-pH soil such as *Gypsophila* and perhaps medicinal plants. On high-quality farmland, which has been used for truffières in both Australia and New Zealand, crops may be grown between the rows. For example, on Alan and Lynley Halls' truffière near Gisborne, New Zealand, which is in a subtropical zone, there was sufficient irrigation water to grow peas, kumara, and melons between the rows in the first few years. This provided some revenue, but perhaps more importantly it raised the organic matter and nitrogen levels in the soil. Alternatively, the land could be left fallow or planted with grass or alfalfa, mown, and the clippings fed to stock. Then, as the root systems of the truffle-infected trees grow out, the grass strip would be gradually reduced. Whatever is done with the land between the rows, it is important that any machinery used does not compact the soil, and any vegetation does not significantly affect the soil's pH. Towards the end of the productive life of a truffière, hazelnuts, firewood from thinnings, and timber from oaks are possible sources of income, but their value is trivial compared to the potential value of the truffles produced over several decades.

When and How to Plant

In Europe planting of truffle-infected host trees is typically carried out in late autumn or early winter (October and November) or spring (late February through early April), traditionally during the waning moon in March. Experience tells that early-winter plantings are superior if there is drought during the following year (not an easy task to predict) and the plants are not irrigated. In New Zealand truffières have been successfully established with both late-autumn and spring plantings. With late-autumn plantings in warm areas of the country, however, there is a danger that dormant plants (the condition plants are normally supplied in) could partially break bud before winter and then be damaged by winter frosts. Worse yet, winter-active competing fungi could become established at a time when the truffle is dormant.

Plants can be raised in a variety of containers including cellulose bags, black polythene bags, square plastic pots, trays with side slots, and Lannen trays. If the plants are in black polythene bags, square plastic pots, and other solid-walled containers, they should be watered and removed from the containers; if the roots have begun to spiral, a sharp knife should be used to make several shallow cuts from top to bottom around the root ball. The trees should then be planted immediately, with the ball of potting soil covered with about 2 cm of field soil and then watered in. To help prevent ponding around the host plant when it rains, the soil should be left slightly elevated (about 2 cm) above the surrounding soil in a circle of 0.5-m diameter. The soil should not be compacted too much around the plant—it is much better to use your hands rather than your feet. Plants supplied in cellulose bags should be soaked for a few minutes in water, the bag cut from top to bottom in three places, while taking care not to damage the roots beneath, and then planted without removing the paper cover, which will soon rot away in the soil.

Some plant producers label and number each plant, which provides some degree of confidence in a buyer that the plants are of a certain standard. By keeping a register of the label numbers along with buyers' contact details and various pieces of technical information, the nurseryman can

then trace where the plants went should he subsequently find that there was a problem with a batch or to later identify those plants that did better than expected. Theft of plants is always a possibility, so keeping the labels in a safe place might later provide a proof of purchase.

Desert Truffles

The desert truffles, in particular *Terfezia claveryi*, are the most recent of the truffles to have been cultivated. Unlike the other truffles, *T. claveryi* grows on rockroses, perennial herbs in the genus *Helianthemum* (Cistaceae). In Spain cultivation has been carried out on low-priced, semi-arid, marginal land. The soils are clay loams with pH around 8.5, 0.9–3.9 per cent organic matter, and a carbon-to-nitrogen ratio between 7 and 10. The infected shrubs are planted at around 1400 plants per hectare. The group led by Mario Honrubia at the University of Murcia produced their first truffles in spring 2001, just under two years after planting. Since then improved methods have brought this down to just 12 months, primarily by irrigating. Although the price for desert truffles is relatively low (€3 to €12 per kilogram), production has been as high as 600 kg/ha, making the short-term returns superior to the other truffles. In 2005, 60,000 rockroses infected with *T. claveryi* were planted in Andalusia.

SIX

Maintenance of Truffières

ANY A PLANT PATHOLOGIST'S AIM in life is to keep plants free of harmful infections or at least to limit them so that the infections do not get out of hand and affect crop production. Towards this aim, farmers plant cultivars that are resistant to pathogens and modify the environmental conditions to make life uncomfortable for the pathogen, such as changing the soil pH and irrigation regime. If all else fails, there is an armoury of chemicals that can be used. For truffle growers, however, the main aim is quite the reverse. They need to keep the host plant infected with the chosen fungus and have the truffle fruiting at the maximum rate, preferably for 50 years or more, while ensuring that the growth of competing ectomycorrhizal fungi is minimized and the plant is free of pests and diseases.

Plant pathologists and farmers have more than 150 years of countless field experiments and a vast investment by chemical companies to call upon. In contrast, truffle growers remain remarkably ignorant of how truffières should be managed, the complex interrelationship each species of fungus has with its host plants, and those environmental factors that are needed to induce fruiting. While science has attempted to rectify the situation,

relatively few long-term experiments have been conducted on the cultivation of truffles. Other research has been limited to monitoring privately owned truffières, such as that conducted in New Zealand since 1985 and Australia since 1993. Understandably, growers have attempted to plug these huge gaps in our scientific knowledge with practices based on traditional beliefs or folklore based on more than 200 years of practical experience. Where these, too, have failed, growers and some scientists have fallen back on intuition, opinion, or outright wild speculation. Although well meaning, these traditional practices almost universally have little scientific basis and at best only relate to truffières in their own region. For example, growers may attribute success to a particular treatment or practice (perhaps planting the truffière by the light of the full moon—and this is not a facetious comment), while the true cause might have been regular rainfall or something else totally unrelated to the attributed treatment or practice and unappreciated by the grower. If a grower is particularly convinced of the efficacy of his or her practices and has a convincing personality, others may be persuaded to copy the procedure or act as its advocate, even though they may have quite different environmental conditions in their truffières.

It is the responsibility of truffle growers to become conversant with the best information available and acquaint themselves with the likely consequences of anything that they do or do not do to their truffière. Likewise, it is the responsibility of scientific advisors to weigh their words carefully and help growers to distinguish between sound scientific fact and what is probable, debatable, a postulate, or wild speculation. Regrettably, this has not always been the case, with enthusiasm sometimes getting in the way. What we have tried to do when writing this chapter is to distil the literature, separate fact from fantasy, present what we believe to be the concrete information, and point out where the boundaries of this information lie.

Each truffle species has its own distinct set of ecological requirements. Factors that have to be considered may include soil pH, moisture, fertility, temperature, aeration, texture, organic matter type and content, as well as canopy cover, including that produced by the host plants, shrubs, and herbaceous vegetation. If a truffle requires a high-pH, free-draining soil, then one cannot expect it to thrive when the host trees completely shade

the ground. Canopy closure will cause the soil moisture level to rise, which in turn may cause the many fallen leaves to rot and drop the pH as they become incorporated into the soil, which will lead to ectomycorrhizal fungi better adapted to the new conditions becoming established. The location of a truffière also has an impact on maintenance. For example, in a cool region a truffle may require an open canopy where the sun warms the ground, whereas in a warm region more shade may be required.

Maintenance Methods

Over the years several management systems for the Périgord black truffle have become recognized, particularly in France. The most primitive is the traditional system in which trees, which may not even be guaranteed as infected with truffle, are planted at 100 to 150 trees per hectare into areas where it is likely that truffle spores will be in the ground. After planting, the owner then simply walks away and essentially leaves the rest to the tender care of nature—there is no working of the ground in spring, no irrigation, and if the trees are pruned it is not intensive. The results are generally uncertain. The Malaurie system is a similar low-budget model, but only 60 to 80 trees are planted per hectare into rejuvenated poor soils of the Causses (limestone plateaus with an elevation of 700–1200 m around the Massif Central, France), with the aim of producing an attractive parkland environment. Production is expected in 15 to 20 years. As its name implies, the renewal system is aimed at resurrecting areas of trees that produced truffles in the past. It involves removing trees, pruning where there is a closed canopy, and working the ground. This system is a risky exercise, as the results are unpredictable, but the costs are not excessive and the potential returns for some can be good.

In the Tanguy system, mycorrhizal seedlings are abandoned once they have become established. The vegetation between the trees, mainly grasses, is allowed to develop naturally, mowed, and the clippings left in place. Using this method, the onset of production generally takes 10 years and some trees may produce very well while others produce nothing. In noncultivated soils, such as in Tanguy-managed truffières in France, truffles tend

to be formed superficially. This has also occurred in a highly productive uncultivated New Zealand truffière on a heavily limed volcanic ash.

The most intensive method is the Pallier system, which was initially devised by Jean Grente and published in 1972 but subsequently embellished. Of the various systems, it is the most expensive and time consuming, but the results are more predictable and fruiting begins earlier. Hence the Pallier system was the one adopted in New Zealand in the 1980s and Australia in the 1990s. It originally required planting 625 to 800 trees per hectare, regular tilling, pruning off the hazelnut suckers, pruning the tree to produce a plant in the shape of an ice cream cone (with the point at the bottom), weed control with chemicals or machinery, and irrigation.

Working the Ground

The Périgord black truffle shares with the Burgundy and Italian white truffles the need for an open-textured, well-aerated soil. There is plenty of evidence that truffles are larger, better shaped, and more numerous in cultivated soils than in those that have never been cultivated. Consequently, it is generally recommended that the soil in a truffière be periodically worked to aerate the ground. Traditionally in France and before the introduction of machinery, this was done using a hand implement, the *bigos*, which looks like a two-, three-, or four-toothed fork but with the prongs at right angles to the shaft. This tool was handled like an adze but only to loosen the soil rather than digging it—a daunting task for a truffière containing more than a few dozen trees. Tractor-mounted sprung tines are now popular in France, Australia, and New Zealand. In France they are often fitted with a rear-mounted roller to prevent the tines dipping below a set depth of 5–10 cm. Another way is to fit a hydraulic ram to one side of the three-point linkage on the tractor that enables the bars carrying the tines to be tilted so that the depth of cultivation next to the trees can be set to 5 cm, while away from the trees the soil can be tilled deeper, typically 8 to 10 cm but never more than 15 cm.

Working the ground seems to be particularly important in the early years of a truffière, before production has begun, and it also helps to con-

Sprung tines are used for incorporating lime or aerating soil. The hydraulic ram on the left three-point linkage allows the tines to be angled so that they do not cultivate as deeply next to the host trees. I. R. Hall, courtesy of P. Nelson and G. Nelson

trol weeds. Cultivating too frequently, however, and particularly with rotary hoes, will eventually lead to a loss of the soil's structure, organic matter, and, during dry spells, the soil itself in high winds. Also, working the ground later than early summer will damage shallow roots, mycorrhizas, and the development of the truffle mycelium in the soil. Consequently, many European truffle specialists now recommend that there should be just one cultivation in spring (late February or March, Northern Hemisphere), although there are some who will cultivate two or three times a year but generally no later than June. The soil should never be cultivated when it is too wet; rather than cultivating a truffle soil during a wet spring, it should be left until the following year.

Brûlé

Plants infected with the Périgord black truffle and Burgundy truffle are marked above the ground by a brûlé, an area around the trees with few

plants. Each year the root system of the truffle host plant goes through a cycle of root elongation and exploration of new soil, mycorrhizal formation, and towards the end of the growing season root dormancy and death of old mycorrhizas. As a consequence, the brûlés grow larger each year. Brûlés normally take several years to develop, although in Alan and Lynley Halls' truffière near Gisborne, New Zealand, brûlés about 0.5 m in diameter had developed by the end of the first summer. Once they have formed, brûlés usually herald the start of truffle production in the following few years. Occasionally, a brûlé may only occur on one side of a tree when other ectomycorrhizal fungi dominate on the other side. Brûlés may also be more obvious at one time of year than another, typically being most prominent in late summer, autumn, and winter. In young truffières during the summer, particularly after rain following a period of drought, they may be so indistinct as to go unnoticed. Sheep's fescue (*Festuca ovina*), stonecrop (*Sedum*), and mouse ear hawkweed (*Hieracium pilosella*) are commonly found in brûlés in Europe, and it has been suggested that sheep's fescue has a beneficial effect on yields because good-quality truffles are often found underneath it.

It is not clear how the brûlé develops, but phytotoxic chemicals produced by the fungus, the destruction of weed roots by the fungus, changes to the soil flora and allelochemicals (substances that suppress the growth of competitors) produced by the mycelium, mycorrhizas, and truffles probably all contribute to its formation. Although some have suggested that a lack of light under the canopy is the cause of the brûlé, in many cases this clearly has nothing to do with their formation. In New Zealand brûlés have also been found around plants infected with bianchetto (*Tuber borchii*), which, like the Périgord black truffle, has been shown to produce allelochemicals. Some competing fungi, including *Scleroderma* and a nonfruiting fungus that French scientists call AD (*angle droit*, referring to the right-angled branching of its hyphae), also produce brûlés. These are frequent competitors in European truffières and may generate much excitement before the real reason for the brûlé becomes clear. Not knowing how the Périgord black truffle produces a brûlé makes it very difficult to guess what bene-

ABOVE A Périgord black truffle (*Tuber melanosporum*) brûlé formed after eight months in Alan and Lynley Halls' truffière, Gisborne, New Zealand. I. R. Hall

LEFT The Périgord black truffle (*Tuber melanosporum*) can form intense brûlés on the holm oak (*Quercus ilex*). I. R. Hall

ABOVE Clearly, shading is not the cause of the brûlé around this Périgord black truffle (*Tuber melanosporum*) infected oak (left of centre), Passo del Furlo on the Flaminian Way, Italy. I. R. Hall

RIGHT Bianchetto truffles (*Tuber borchii*) are not supposed to produce brûlés, but these seem to be forming in some artificial truffières in New Zealand. I. R. Hall

This brûlé produced by the contaminant AD ectomycorrhizal fungus will never produce truffles. I. R. Hall

fit it may have to the fungus. If the fungus gains nutrients from the dying weed roots, that would be sufficient impetus for the fungus going down that particular evolutionary path. A lack of competition from weed species, no doubt, would be of assistance to the host plant as well.

Opinion is divided on whether cultivation should be carried out inside the brûlé after production has begun. In many large truffières in France, the brûlés are cultivated in spring (late February or March) with the same machinery as used between the rows but with the tines set to only 5 cm. The brûlés are not cultivated later than early April (Northern Hemisphere), as this will disrupt the developing truffle mycelium and kill the truffle primordia.

Herbicides Versus Soil Cultivation

In France, Australia, and New Zealand weeds in truffières are often controlled with herbicides such as glyphosate, a systemic herbicide that is readily absorbed by leaves and stems and translocated throughout the plant, or the nonsystemic desiccant herbicide glufosinate-ammonium. (Note that desiccant herbicides containing a mixture of 2,4-D and dichlorprop should *not* be used.) These can be applied with, for example, a backpack sprayer or motorized units mounted on a tractor or a four-wheeled farm vehicle. In young truffières, the use of rigid tree protectors enables herbicides to be used quite close to the trees, leaving only a few weeds inside the guards to be removed by hand. Glyphosate can be used to control difficult grasses such as couch (or creeping twitch, *Agropyron repens*), although three or four applications may be needed during a season to finally bring it under control. Glufosinate-ammonium can also be sprayed onto hazelnuts with a woody trunk to kill suckers. It is also effective in controlling stoloniferous clovers, which can completely colonize brûlés.

Because repeated use of glyphosate selects against grasses in favour of dicotyledons, Sourzat (1989) warned against overuse of this herbicide. Instead, he suggested working the ground in spring, when it will not have any detrimental effects on truffle production, and then controlling weeds by mowing. However, in areas with relatively high rainfall, tilling the soil alone is not sufficient to control grasses and weeds, which can easily get out of hand. Continuous mowing will also return fresh organic matter to the soil and drop the soil pH or at least expose the mycorrhizas to acidic percolates. Also, leaving a carpet of grass around the trees will drop the average soil temperature by several degrees, a potential problem in areas that are climatically marginal for truffles, and minimize fluctuations in soil temperature that are normal in natural brûlés and have been suggested as likely triggers to fruiting. It would certainly be unwise to aerate the ground or even mow the grass when fully developed immature truffles may be present just under the soil surface—as early as late February in New Zealand (the equivalent of late August in the Northern Hemisphere).

Whether herbicides have any direct effect on ectomycorrhizal fungi is

A four-wheeled farm vehicle with a rear-mounted spray unit that is used for the control of weeds and suckers on hazelnuts. I. R. Hall

not clear. In one experiment the herbicides triclorpyr, imazapyr, and sulphometuron methyl had no effect on the capability of ectomycorrhizal fungi to infect roots, even at concentrations detrimental to seedling growth, whereas other experiments have shown that glyphosate actually stimulated mycorrhizal formation. Herbicides can have adverse effects on other edible ectomycorrhizal fungi, although it has not been determined whether this is a direct effect of the chemical on the fungus or the denuded ground getting hotter and the soil drier in the sun.

Bumper harvests have been observed following aeration to 30 cm in spring, perhaps because the truffle fungus is able to tap nutrients in roots that have been damaged. Long-term deep aeration of the ground will be detrimental, however, as it will damage the feeder roots and mycorrhizas that are usually in the top 10 to 20 cm of soil.

Soil Moisture

The highly aerated nature of Périgord black and Italian white truffle soils ensures that immediately after one of the sporadic, heavy thunderstorms

that punctuate summer in southern France and Italy, the air within the soil is rapidly replaced by heavier, cooler, oxygen-rich air from above the soil, which has been suggested as being a possible trigger for fruiting in the Italian white truffle. Yields are certainly highest when a moist autumn is followed by a winter with many freeze-thaw cycles and a summer with many thunderstorms and soil temperatures not exceeding 20°c. The sudden 5°c drop in 10-cm soil temperature seen in *Tuber melanosporum* truffières in France has also been observed in New Zealand just a few weeks before the first immature truffles can be unearthed; this temperature decline has been suggested as a possible trigger for fruiting. Similarly, declines in soil temperature and increasing soil moisture have been suggested as being involved in the fruiting of porcini (*Boletus edulis*), another edible ectomycorrhizal mushroom. The answer to what triggers truffle fruiting is almost certainly captured within the paper by Bardet and Fresquet (1995; see also Appendix 9), which examined the timing of certain climatic features in relation to fruiting. While we are still left guessing how we might go about influencing yields, soil moisture and irrigation have been shown to affect both fruiting and yields, and some additional likely candidates are listed in Appendix 10.

Although summer moisture is a requirement for fruiting by the Périgord black truffle, its ability to survive in dry, well-aerated soils with widely ranging soil temperature may well be an ecological strategy that enables

Very immature Périgord black truffles (*Tuber melanosporum*) found in March (late summer) in Alan and Lynley Halls' truffière in New Zealand seem to be associated with a decline in soil temperature of about 5°C. I. R. Hall

it to dominate in a soil where other ectomycorrhizal fungi cannot. Thus, there is a danger that excess soil moisture year after year will reduce the soil porosity, aeration, and soil temperature fluctuations, while leaves and other organic debris will rot too quickly in the soil and lower the soil pH, and that together these factors could alter the ecological balance and allow the entry of competing ectomycorrhizal fungi such as *Scleroderma*. Partly for this reason, some argue that fallen leaves should be removed from a truffière before they alter the soil conditions, a concept that goes back to the late 19th century (Baring-Gould 1894): "They must have freedom in which to develop; moreover, what kills them is the accumulation of dead leaves, or any substance above them, which excludes light and air. An excellent truffle ground has been ruined for years by the accumulation on it of faggots that have been left, and not immediately removed. Moreover, much injury is done in an oak wood when the trees are felled, by dragging the timber along the soil, as it tears up the tubers, and injures the fibrous roots on which they feed."

Many trees, and in particular poplars, are renowned for sending their roots out by as much as 30 m in search of water. If these trees are adjacent to an irrigated truffière, they could not only compete for water but could also carry contaminating ectomycorrhizal fungi into the truffière. One way of controlling this problem would be to cut the trees down, but this might not be a good idea if they are on a neighbour's property. An alternative is to put a trenching machine through the soil between the truffière and the neighbouring trees. Typically, these cut a slot in the soil 10 cm wide and down to 90 cm and will successfully sever even quite large roots. This will need to be repeated on a regular basis to stop any regrowth of the roots.

Irrigation

Truffle production is severely affected in dry years and it is obvious that adequate soil moisture, as supplied by summer thunderstorms in Mediterranean climates, is essential for fruiting. Streams and rivers are not a common sight in the limestone-rich regions of southern Europe, but where there is irrigation water, growers may use an intuitive approach and irrigate when the soil looks as though it needs it. Taking a handful of soil collected

from a few centimetres below the soil surface and squeezing it is a quick and easy way of determining whether the soil is dry. Others follow rather general recommendations for irrigating: "after about 20 days of drought, 30 mm of water is to be applied" (Sourzat 1981). Some truffle growers use tensiometers (devices for measuring the moisture status of a soil) and irrigate when the gauge registers more than about 40 centibars, while a few have installed electronic moisture sensors that trigger sophisticated irrigation systems. The problem with these is they rely on the gypsum block at the bottom of tensiometers or the electronic sensors to be in intimate contact with the surrounding soil, something that requires careful maintenance. Whatever system is used, when irrigating it is important that the rate at which water is applied does not exceed the infiltration rate, as this will result in ponding and, particularly in soils with a relatively high clay content, a loss of the soil porosity known to be essential for the Périgord black, Burgundy, and Italian white truffles.

One might have thought that our lack of knowledge of what triggers the fruiting of edible ectomycorrhizal mushrooms and the factors controlling the size of the harvest would have spawned a rash of field-based research, at least for those species that can be cultivated. However, publications on this topic are rare, so it remains a fertile area of research for someone with 20 years to spare—and the research funding to go with it. Hopefully, with irrigation becoming more popular in European truffières based on some spectacular experimental results and with irrigation being almost mandatory in the drier parts of Australia and New Zealand, more reliable information on how much water should be applied and when will become available over the next decade.

Mulching

In the truffle-growing areas of Europe where irrigation water is either unavailable or simply too expensive to use, growers have traditionally placed mulches such as straw, leaves, juniper branches, or, more recently, polythene film over the brûlé. The short-term effects of mulching include increased soil moisture and equilibration of soil temperature, whereas long-term responses are increases in soil mineral nutrients, soil aggrega-

tion, microbial activity, and enzyme activity and improved drainage characteristics. As a consequence, mulching can deter the development of diseases, insects, nematodes, and soil-borne fungi and thus have a positive effect on plant growth and truffle production. However, mulches have been shown to reduce soil temperatures and soil degree-days, which may have a detrimental effect in truffières in areas with climates less suitable for truffle production. Also, straw and black cloth mulches placed around young truffle-infected plants have been shown to depress truffle infections and stimulate competing ectomycorrhizal fungi, which brings into question what the long-term consequences might be of the use of mulches in truffières.

Fertilizers

In Europe several specialized organic fertilizers for truffières are used to raise the organic matter levels in the soil or to mulch the brûlés. These are composted mixtures of waste materials of plant and animal origin adjusted to above pH 8.0. One of these, Fructitruf, a mixture of bones, skins, castor oil seed waste, and potassium hydroxide, has been applied at 1 kg/m^2 around young trees or at 500–3000 kg/ha in older truffières. While the incorporation of rice husks at 250 g/m^2 of brûlé has been reported to increase the average number and weight of truffles in France, this has not become standard practice.

Phosphatic fertilizers generally lower the amount of mycorrhizal fungi on the roots of plants. Also, the most productive New Zealand truffière has a low concentration of plant-available phosphorus. Consequently, in New Zealand the application of phosphatic fertilizers and limestone with a high phosphate concentration (>0.1 per cent) has been discouraged. However, even high applications of phosphatic fertilizers in French truffières have had no significant effect on the Périgord black truffle. Similarly, in the New Zealand truffière owned by Don and Isabel Dempsey, truffles have formed and there have been good infection levels even with 70–110 µg/ml of Olsen extractable phosphorus, a level more akin to a horticultural soil. Clearly, our understanding of what really happens under the ground in truffières leaves quite a lot to be desired.

Nutrient Deficiency

While soil analysis is a very useful tool in measuring how much of the various elements are present, it is only an approximation of how much is available to the plant and how much the plant will absorb. It is better to measure what is inside the plant, particularly in the leaves, the powerhouse of green plants. Because foliar nutrient concentrations change as the leaves age, samples of the foliage of hardwood trees are normally taken in late summer and well before leaf fall (late August to September in the Northern Hemisphere), while for softwood trees sampling should be done after the trees have gone dormant in late autumn. To minimize errors, about 10 plants need to be sampled. These can be chosen randomly, along a line drawn through the truffière, or just from the plants that seem to be affected by nutrient deficiencies. Separate samples need to be taken where there are several host plants in a truffière because the normal healthy nutrient concentrations vary from species to species. Plants in dusty areas, such as next to roadsides, should not be sampled because the quantities of some elements in the dust could be much higher than what are inside the plants. If there are few plants in a truffière or the plants are very small and sampling would remove a significant proportion of the foliage, it is better to analyze the soil rather than stress the plants.

When collecting foliar samples it is important that they be collected into new, clean paper bags. Polythene bags should not be used as they make the leaves sweat and respire, which can lead to inaccuracies in the analyses. It is best not to touch the leaves with sweaty hands as this can also cause inaccuracies as well. For broad-leaved species only the uppermost, fully expanded mature leaves should be sampled, and for softwoods only leaves from the growing season just finished. A total of 50 g fresh weight (about 5 g after drying, or 20 to 30 leaves of large-leaved plants) is usually more than adequate, but it is best to first check with the laboratory you will be using. The leaves will lose weight and deteriorate in the mail, so if you are more than 24 hours from a suitable laboratory you will need to dry the leaves at 70°c for about 24 hours. This can be done by putting the leaves on clean upturned lids of glass casserole dishes in a domestic oven set on fan

bake. To be sure that the oven temperature is within ± 5°c, it can be calibrated with an inexpensive digital thermometer with an external probe purchased from your local electronics store. Specialist laboratories dealing with foliar analyses can be found via the Internet or a telephone directory. A very small selection of these can be found in the back of this book. The results of the analyses can then be compared against standards. Normal nutrient concentrations for hazelnut trees grown for their nuts are available from a variety of sources, but standards for species such as hop hornbeam, holm oak, and rockrose seem to be unavailable.

Potassium, magnesium, and sulphur

Potassium deficiency can be corrected by the application of potassium sulphate at 100 kg/ha. Those involved in the fruit-tree industry, however, caution against the application of potassium on its own because of the risks of unbalancing the potassium-to-magnesium ratio. The recommendation is therefore to apply sulphur-coated potassium sulphate at 1–4 kg/ha per tree and about one-third by weight of controlled-release magnesium sulphate. Then, once the potassium deficiency has been corrected, no more potassium should be applied until foliar analyses shows it is again needed.

A magnesium deficiency in a soil is likely to have been detected in the preliminary soil testing conducted well before a truffière was established and corrected by the incorporation of dolomite into the soil. If a magnesium deficiency develops after establishment, however, it can be treated by the application of fast-acting magnesium sulphate at about 100 kg/ha or by the application of finely ground dolomite. Sulphur deficiency can be corrected by the application of fine elemental sulphur at 30 kg/ha.

Trace elements

In high-pH soils some trace elements that are essential for plant growth can be locked up in insoluble forms that plants cannot access, causing symptoms of trace element deficiency. This also happens when a naturally acidic soil is heavily limed. Iron is the element most likely to be affected and cause problems in high-pH soils, although boron, copper, manganese, and zinc deficiencies also may occur. The main symptom of trace element deficien-

cies is yellowing (chlorosis) of the foliage. Generally, deficiency symptoms of each of the trace elements are similar, particularly when the deficiencies are just at the threshold. Even the experienced eye will temper a diagnosis with a degree of caution until foliar analyses have been obtained.

Iron

When the available soil iron concentrations are below about 100 ppm, this can lead to iron deficiency problems in plants. Characteristically, these appear as yellow leaves (chlorosis) with green midveins or in more severe cases as browning of the leaves and severe disruption of growth. In limed acidic soils, iron deficiency symptoms may disappear spontaneously once the plant roots grow into the subsoil, where the available iron concentration is often higher. However, in naturally high-pH soils the available iron concentrations can be even lower in the subsoil because of large amounts of limestone from the underlying rock.

In the hazelnut industry, iron deficiency is remedied by applying 100 g Fe-EDDHA (iron ethylene diamine dihydroxy acetate) to the soil under each tree at the end of winter. Iron deficiency in young truffle plants can be alleviated by the application of 15–50 g per tree of an Fe-EDDHA chelate to the soil inside a 0.5-m-diameter circle around each tree. Because iron chelates are light sensitive, it is necessary to bury them beneath the soil surface to prevent deterioration. This can be achieved by raking 1–2 cm of soil from around the trees, sprinkling on the chelate, and then replacing the raked soil. The alternative is to apply regular foliar sprays during the growing season—a cheaper alternative but only in chemicals, not if your time and back are taken into account. Please note that iron chelates are considered to be harmful if swallowed or inhaled. They may also cause irritation of the eyes, nose, throat, or skin, so contact should be avoided. If you get any of the chemicals in your eyes, they should be immediately washed with plenty of clean water, and then a doctor should be consulted.

Manganese

Low levels of extractable soil manganese are common in high-pH soils and where large amounts of lime have been applied to a soil, as found in many New Zealand truffières. Manganese deficiency symptoms are often simi-

lar to those for iron deficiency, but with darker green veins separated by yellowish areas (interveinal chlorosis). Although iron deficiency tends to affect younger leaves and manganese deficiency is more pronounced on leaves of intermediate age, this is not an easy or reliable characteristic to use as a diagnostic tool. Instead, it is better to have foliar analyses performed and have these interpreted by a specialist. Manganese deficiency symptoms in most species are associated with leaf levels less than 20 ppm, with particularly severe symptoms at less than 10 ppm. Healthy plants normally contain 50 to 200 ppm of manganese, although levels up to 1500 ppm have been recorded where fungicides containing manganese have been applied.

Manganese deficiency can be treated by spraying the foliage with manganese sulphate at 5–30 kg/ha, or at a concentration of 1–5 kg of manganese in 200 L of water. However, with manganese-deficient citrus the plants have to be sprayed every week. Instead of foliar sprays, powdered manganese oxide, manganese chelate, or manganese oxysulphates can be applied to the soil (Appendix 11). For a broadcast application, manganese oxysulphate should be applied at 50 kg/ha; if applied in a furrow or band, 10–20 kg/ha should be sufficient. Similarly, if the manganese is to be sprinkled around each tree, the quantity applied should be no more than 2 g/m². Because broadcast manganese should only have to be applied once a year, the small additional cost in chemicals would certainly be cheaper than the cost of spraying the foliage every week.

When manganese and iron are applied simultaneously, they can interfere with the uptake of each other. If you are in the unhappy situation where both are needed, we recommend that you seek professional advice.

Copper

In hazelnuts and oaks, the major feature of severe copper deficiency has been referred to as "dieback" or "witches broom" because of the early cessation of the growth of the terminal bud and partial dieback of terminal buds. As with the other trace elements, the main symptom of marginal copper deficiency is chlorosis. High concentrations of iron, phosphorus, or zinc in the soil solution can depress the absorption of copper by plants and induce copper deficiency.

Because many copper compounds are fungicidal, considerable care needs to be taken when applying copper to a truffière—sufficient copper needs to be applied to meet the plant's needs but without it having a detrimental effect on the fungus. High copper concentrations can also have a detrimental effect on root growth. A conservative approach would be to apply copper sulphate at 1 kg/ha in the irrigation water, with a concentration no greater than 0.25 kg per 1000 L of water. This could be repeated each irrigation until the foliar copper concentrations approach the normal range. An alternative is to broadcast the dry chemical directly to the soil surface and then irrigate it in. However, it is difficult to apply very small quantities of fertilizers over large areas, so trace element fertilizers are often mixed with fine dry sand and the mixture spread. One way to ensure an even spread of this mixture is to combine 1 kg of copper sulphate with 20 kg of fine dry sand for each hectare that needs to be treated. This mixture is then divided into 16 parts, and 1 part of the sand/copper sulphate mixture is then spread over $1/16$th of the truffière.

In Europe truffières are often established in areas that have been used for crops in the past, in particular vineyards. In these areas large quantities of the copper-containing Bordeaux mixture have generally been used in the past to control pathogenic fungi. Consequently, the concentrations of copper in the soils can rise to the point where they are toxic to plants and copper may accumulate in truffles.

Boron

The symptoms of boron deficiency include a general yellowing of the leaves, with young leaves more affected than older ones; small stunted plants with rosettes of misshapen, small, thick leaves on the tips of branches; and death of the branch tips. Boron deficiency tends to be exaggerated by drought.

In the hazelnut industry of North America, a single application to each tree of 125 g of fine sand containing 0.1 per cent boric acid or borax (that is, 0.125 g/m^2 mixed with 125 g of sand) reduces the proportion of empty nuts, increases the size of nuts and total yields, and encourages the growth of primary and lateral roots. Boron deficiency in New Zealand's *Pinus radiata* forests is typically controlled by broadcasting finely ground ulexite chips at

60 kg/ha (8 kg/ha of boron) one to two years after planting. Sodium borate applied at a rate of 2 kg of boron per hectare will a give rapid recovery if dieback is already occurring (Appendix 12).

The amount of boron needed by plants is very small, and it is relatively easy to apply too much, resulting in boron toxicity. Consequently, rates not greater than 0.5 to 1.5 kg/ha of boron are normally recommended for crops. In the hazelnut industry, it is recommended that no additional boron be applied if the foliar boron concentrations rise above 200 ppm.

Zinc

We have not yet identified zinc deficiency in truffières, but we are sure it must exist. Zinc deficiency is common where crops are grown on sandy, calcareous, or sodium-rich soils, in which there are generally low levels of zinc. Deficiency also occurs where there are high levels of soil phosphate and nitrogen, the roots are restricted through soil compaction, or there is a high water table. The symptoms of zinc deficiency in crops and fruit trees are yellowing of the parts of the leaves between the veins and the formation of rosettes of leaves at the tips of each stem.

Applications of zinc sulphate are commonly used for treating zinc deficiency, either by applying it to the soil or as a foliar spray, although including it in irrigation water seems acceptable. Broadcast applications of 11 to 100 kg/ha of hydrous zinc sulphate or 0.3 to 6 kg/ha of chelated zinc are typical *in agriculture*. However, like copper, zinc is fungicidal and so we recommend using 1 kg/ha in the irrigation water, with a concentration no greater than 0.25 kg per 1000 L of water. The residual effect of zinc treatments can last several years.

Plant Diseases

Ectomycorrhizal fungi are able to access some nutrients directly from the soil, but they are largely dependent on the host plant for carbohydrates and a place to live. Plant diseases caused by fungi, bacteria, and viruses and invertebrate pests rob the plant of its resources, may disrupt photosynthesis, and interfere with the plant's ability to translocate nutrients from the roots to the

shoots and vice versa. Consequently, it is logical that truffle growers should do what they can to maintain the health of their host trees.

Many diseases and pests can affect truffle hosts. For instance, oaks in the United States are home to more than 90 bacterial and fungal diseases and 100 pests, all capable of obtaining nutrition directly from the trees. Some of these have the capacity to seriously threaten the health or even the very survival of trees and hence the productivity of a truffière. Consequently, there is a fair chance that the owner of a truffière will experience some problems from one or more diseases or pests.

In some cases the cause of a disease is not due to another organism but abiotic factors, such as frost, excessively high soil or air temperatures, drought, pollution, or herbicides. Preventing these by avoiding drought or overwatering, the careful use of herbicides, and measures such as maintaining the growing conditions in a state that do not favour disease organisms are often simple to achieve. Similarly, preventing diseases from occurring is better than trying to cure them or stop them from spreading once they have entered a truffière. For example, planting only certified disease-free host plants, using tools that have been thoroughly washed and soaked for a couple of minutes in a 5 per cent solution of domestic bleach, and painting cut surfaces with pruning paste or even acrylic paint may be all that is required to prevent the entry and spread of a disease.

Many specialist books on plant diseases are available, such as *The European Handbook of Plant Diseases* and the excellent series of books and Web pages published by the American Phytopathological Society. Therefore, here we describe just a few examples of the problems a truffle grower might encounter.

Fungal diseases

Several fungal diseases can infect truffle-infected plants, including foliar spots, rusts, cankers, root rots, collar rots, and wood rots. When preventative measures are taken, most do not cause severe problems in the field, but some are troublesome—particularly inside greenhouses, where the elevated temperatures and humidity favour growth and spread of fungal diseases. In truffières avoiding irrigating during the hottest hours of the day

and providing plenty of space for air movement between trees are measures that will limit the spread of a fungus.

Fungicides can be used to try to control these diseases, but care has to be taken to ensure that the fungicide does not run down onto the soil and possibly harm the truffle fungus. Many fungicides are active against Ascomycetes, the most common pathogenic fungi, and consequently also inhibit truffle mycelium more than ectomycorrhizal competing fungi, which are mostly Basidiomycetes. Many Web sites can be used to select the best fungicide to use according to their activity, formulation, and application (for example, the Italian site http://fitogest.imagelinenetwork.com). However, practical advice can often be found by entering, for example, "fungicides" and "oak" or "hazelnut" into a search engine on the Internet.

The use of fungicides and other potentially poisonous compounds in the cultivation of a very highly priced gourmet product like truffles obviously has to be carefully considered. Fungicides certainly cannot be used if it is one's intention to produce truffles free of agricultural chemicals, but their judicious use where the survival of the host plant might be at stake is certainly warranted. While some researchers have found that the fungicides iprodione and maneb had no depressive effects on mycorrhizal formation, or even stimulated it, others have demonstrated that iprodione suppressed bianchetto truffle in favour of *Hebeloma*, an ectomycorrhizal Basidiomycete fungus.

Powdery mildews (Erysiphales) are very common fungal diseases of a wide range of plants. These fungi include *Microsphaera alphitoides* (= *Oidium quercinum*, the asexual stage), *Microsphaera alni*, *Erysiphe trina*, *Phyllactinia corylea*, and *Sphaerotheca lanestris* on oaks and *Phyllactinia guttata* on hazelnuts. The disease is easily recognized from the white to greyish powdery spots or patches on leaves, young stems, buds, flowers, and young fruit, particularly when there is hot and humid weather. In severe cases, the leaves and shoot tips can be covered by the fungus, which can lead to necrosis and partial or total defoliation.

Ergosterol biosynthesis inhibitors (EBIs) or elemental sulphur are recommended for the control of powdery mildew. Fortunately, EBIs and sulphur have been shown to have no observable effects on bianchetto and Bur-

gundy truffle ectomycorrhizas, even at high rates of application. There is also a biological control agent that is effective against powdery mildew, *Ampelomyces quisqualis*, a naturally occurring hyperparasitic fungus. It infects the powdery mildew hyphae, conidiophores (specialized spore-producing hyphae), and cleistothecia (spherical and closed fruiting bodies) and reduces growth of the mildew colony. A powder containing *A. quisqualis* is available with the trade name AQ10.

A nontoxic measure to prevent powdery mildew is to mix 1 tablespoon baking soda, 2½ tablespoons olive oil, and a few drops of dishwashing detergent with 5 L of water, shake well, and spray onto the leaves. The baking soda is fungicidal and the olive oil puts a greasy film on the surface of the leaves and inhibits the fungal spores from germinating and penetrating through the oily surface and into the leaves.

Eastern filbert blight, caused by the Ascomycete *Anisogramma anomala*, is one of the main diseases to affect the European hazelnut (Corylus avellana) in North America. This fungus causes minor problems on the American hazelnut (*Corylus americana*) but is a potentially fatal disease on the European hazelnut. The Pacific Northwest was free of this disease until 1973, but eastern filbert blight is now widespread throughout North America. The first symptom of the disease is the appearance of raised black cankers (stromata) about 12 months after infection. These rapidly spread around the tree by as much as 1 m per year. Withered leaves remain attached to the dying branches and are characteristic of the disease. Without treatment, well-established trees will die within 5 to 10 years. In the North American hazelnut industry, control measures involve removing infected branches 0.5 to 1 m below where the symptoms appear, burning them, and then spraying the trees with compounds containing copper hydroxide. Resistant cultivars and pollinators are also recommended in the hazelnut industry to limit the spread of the disease.

Phytophthora is a genus of aquatic, amphibious, and terrestrial fungus-like organisms with motile spores in the large primitive family Pythiaceae that cause serious diseases of economically important plants (for example, timber and ornamental trees) and crops. *Phytophthora infestans* was responsible for the potato famine in Ireland beginning in 1845, and in Australia

P. cinnamomi has caused severe dieback in a variety of trees in native forests. *Phytophthora ramorum* is the cause of sudden oak death in the United States and has spread to Europe. *Phytophthora ramorum* and *P. kernoviae*, another species recently found in Europe, are not yet a problem in the truffle industry in Europe and North America but have the potential of becoming so: particularly susceptible trees include Turkey oak (*Quercus cerris*), holm oak (*Q. ilex*), European beech (*Fagus sylvatica*), and Douglas fir (*Pseudotsuga menziesii*), which are all major host plants for truffles (Appendix 3). As with other diseases, those truffières containing mixed plantings of several unrelated species of host plant are likely to fare better than those containing just one susceptible species. For example, holm oak is more prone to infection by *P. ramorum* than is the common oak (*Q. robur*).

Good soil drainage can prevent or limit the spread of root and collar rots, cankers, and the decline and death of plants caused by species of *Phytophthora*. Removing infected trees or pruning off infected branches and then burning them followed by spraying the pruning wounds and surrounding trees with copper formulations will help. However, it is better to prevent these diseases than to try to cure or limit their spread after they have entered a truffière.

In Spain and Italy *Cytospora corylicola*, *Nectria ditissima*, and *Sphaceloma coryli* cause canker disease on hazelnuts, which results in the death of older branches. Other common species that can cause cankers on oaks in Europe include *Nectria galligena*, *Nectria cinnabarina*, *Hypoxylon mediterraneum*, and *Botryosphaeria*. Their treatment and control is similar to those used for Eastern filbert blight.

Root rots caused by *Armillaria* species and *Rosellinia necatrix* are common in Europe and elsewhere and are able to kill trees or shrubs, particularly when plants are under stress. Because these fungi develop on the roots, they often remain undetected until their characteristic mushrooms are produced around the base of the trunk. Infected plants appear unhealthy with sparse, discoloured foliage turning yellow and then brown and show a dieback of the branches. These very dangerous diseases are better avoided by not planting truffières in recently cleared forests. In addition, *Armillaria* can rapidly spread through the soil in ropes of fungal hyphae (rhizo-

morphs), which can be several metres long, so it is essential to avoid planting near plants infected with this fungus. Poorly drained soils lacking oxygen are also best avoided because these conditions can weaken plants and make them more susceptible to root rots.

Rusts are pathogenic fungi that primarily grow on the stems and leaves of plants and produce patches of rusty-coloured spores. During the production of truffle-infected pines it is sometimes necessary to apply systemic fungicides to control rusts, but there is a risk that they will affect the development of truffles on the roots. Several fungi antagonistic to or parasitic on rusts are known, but these have not yet been developed as biological control agents.

Bacterial diseases

Truffières are usually not affected by bacterial diseases, although in the hazelnut industries of France, southern Italy, eastern Europe, and Oregon *Xanthomonas campestris* pv. *corylina* can cause severe damage. This bacteria causes roughly circular necrotic spots, sometimes surrounded by a yellowish halo, first on buds and then on branches and leaves arising from the bud. The disease can sometimes advance to produce dark soaked cankers on the shoots and may completely encircle and kill young trees. Lesions on stems can ooze bacterial cells in a sticky liquid that is capable of infecting other plants that have been physically injured.

Another bacterial disease that produces a severe blight on hazelnut is caused by *Pseudomonas syringae* pv. *avellana*. This currently affects about 1000 ha of hazelnut groves in central Italy alone. The disease can be detected in spring, when infected plants show rapid wilting because the bacteria interfere with water translocation within the plant.

Removing and burning dead branches and badly affected trees and using copper-based chemicals are partially successful in controlling the spread of bacterial diseases, particularly when there are pruning wounds or damage caused by spring frosts, hail, and wind and at the beginning and middle of leaf fall.

Viral diseases

The most significant viral disease is hazelnut mosaic, caused by the apple mosaic virus (ApMV). ApMV has been found in a large number of hazelnut cultivars from Spain, southern Italy, and Turkey. Symptoms include chlorotic spots, striations, and flecking on older foliage. The virus can be readily detected in young foliage with an ELISA test. Essentially nothing can be done to treat viral diseases except planting virus-free trees to begin with—another problem that nurserymen need to be aware of.

Invertebrate Pests

Host trees in truffières can be damaged by a wide range of insects that attack the leaves, roots, or wood. The larvae of the European goat moth (*Cossus cossus*) and wood moth borer (leopard moth, *Zeuzera pyrina*) can burrow into the wood and are particularly troublesome in southern Europe. The wood moth borer is the most dangerous because the larvae can kill young plants. The most effective way of controlling this pest is to use semiochemical traps. While the insecticide teflubenzuron can be applied to the trunks to prevent the adults laying eggs (spring to summer for the wood moth borer and summer for the European goat moth), quite high concentrations have to be used and it is toxic. Another highly toxic pesticide, aldicarb, is widely used in pecan orchards in North America; consequently, people are warned against harvesting pecan truffles in these areas. Strangely, the insect *Cerambyx cerdo*, which can cause severe damage to adult oaks, particularly in Marche and Tuscany, is protected throughout the European Union.

Sciarid flies

Breeding in warm moist places where there is decaying organic matter, sciarid flies can be a major problem in propagating greenhouses. The adults lay their eggs just under the surface of the potting mix, and the tiny larvae feed on the developing roots. The adult flies are attracted to certain colours, and hanging yellow or blue sticky pads around a greenhouse can provide an effective control without resorting to the use of pesticides such

as diflubenzuron. A mixture of 1 tablespoon baking soda, 2½ tablespoons olive oil, and a few drops of dishwashing detergent in 5 L of water sprayed on the leaves seems to help to control sciarid flies in greenhouses where truffle-infected plants are produced. The adult flies stick to the oily surfaces, and this helps to break the breeding cycle. Parasitic nematodes can also be used. If sciarid fly infestations get really bad, the insecticides pirimicarb and diflubenzuron can be used. As with all agricultural chemicals, these should be applied with caution.

Black beetle

In New Zealand, African black beetle (*Heteronychus arator*) adults have caused severe problems in newly planted truffières. In Northland the adult beetles burrowed their way into the root mass inside the cellulose bags of newly planted oaks and hazelnuts and then proceeded to strip the outer layers off the lower section of stem down as far as the phloem—the vascular tissue that transports photosynthates around a plant. Most of the smaller affected hazelnuts (about 25 cm high) were quickly killed, whereas larger hazelnuts with thicker bark on the lower stem and most oaks, which presumably did not suit the beetles' palate, were not as severely damaged. While such damage is likely to be sporadic, related to season, and restricted to the upper part of the North Island (34°s to 39°s) in sandy, peaty, and free-draining loams, this is cold comfort for the owners of an affected truffière who might have lost a third of their plants.

Because adult black beetles are only around for a few months between spring and early summer, simply delaying planting until midsummer can avoid the damage caused by them. However, planting at the height of summer can have its own problems, and an efficient irrigation system is essential. Another option is to limit the number of black beetle larvae by frequently working the soil in the autumn and winter before planting, which may physically damage the larvae or bring them to the soil surface where they may die of exposure. However, this does not seem to be a reliable control measure. Chemical sprays may also be used to control both larvae and adult black beetles. The systemic insecticide imidochloprid can be used to protect the host trees during the establishment phase or applied as a

drench to the soil at planting, with repeat applications made every 12–14 weeks when risk is highest.

Other pests

Other significant insect pests on hazelnut trees include aphids, in particular *Myzocallis coryli* and *Corylobium avellanae*; filbert worm (*Melissopus latiferreanus*), which is a problem in Oregon; brown gooseberry scale (*Eulecanium tiliae*); European leaf roller (*Archips rosanus*), which can also infest poplars and willows; the southern green stink bug (*Nezara viridula*); and

The larvae of the wood wasp (*Sirex noctilio*), one of the many insects that can cause serious damage to oaks, have almost girdled the trunk of this tree. The only signs of their presence are a few holes and sunken areas around them. I. R. Hall

The big bud mite (*Phytoptus avellanae*) produces these swollen buds on hazelnuts, which at times can reach epidemic proportions and disrupt the growth of the tree. I. R. Hall

the almost ubiquitous big bud mite (*Phytoptus avellanae*). In Europe more than 100 insect pests can be found on the leaves of oaks, such as the European oak leaf roller (*Tortrix viridana*); the European gypsy moth (*Lymantria dispar*), which can defoliate huge areas of forest; oak procession moth (*Thaumetopoea processionea*); lackey moth (*Malacosoma neustria*); white satin moth (*Leucoma salicis*), although this is found primarily on poplars and willows; tussock moth (*Orgya antiqua*), which is also found on willows and various shrubs; and several wood wasps such as *Sirex noctilio*. In the United States there are many more insect pests, all of which may cause problems from time to time and for which local advice on their control may be needed. A good general control measure against many leaf-eating caterpillars is a spray containing the insect-killing bacterium *Bacillus thuringensis*, which is active against the gypsy moth (*L. dispar*), spruce budworm (*Choristoneura occidentalis*), jack pine budworm (*Choristoneura pinus pinus*), hemlock looper (*Lambdina fiscellaria fiscellaria*), tussock moth (*Orgyia thellina*), tent caterpillar (*Malacosoma americanum*), pine processionary moth (*Thaumetopoea pityocampa*), and others.

Because of isolation and vigilant border controls, there are far fewer pests and diseases of Northern Hemisphere host plants in countries of the Southern Hemisphere, including Australia, Chile, New Zealand, and South Africa. The main insect problems in New Zealand truffières have been leaf-rolling caterpillars and aphids, which have been controlled with contact insecticides such as malathion or for light infestations simply nipping the caterpillars with a thumb and finger. Grass grub beetles (*Costelytra zealandica*) have also caused some damage to plants in November and December. They can be controlled by spraying the trees with insecticides such as synthetic pyrethroids or organophosphates—the latter usually being very toxic to people, however.

For young trees in narrow plant protectors, it is very important just how much volatile insecticide is applied, as the concentration inside will be higher than would normally be the case around plants out in the open air. This will be particularly true on hot days when the concentration of the insecticide could become phytotoxic as well as insecticidal. The general rule is to apply the chemical at the recommended rate but only to a few

plants, wait a few days to see if there is any wilting of the foliage or other damage, and then, if all is well, spray the rest of the trees.

Truffle Pests, Diseases, and Browsers

Truffles can also be affected by pests, diseases, and browsing by large and small animals. Insect damage is common in Europe during spring, summer, and autumn when high temperatures favour insect activity. The small truffle beetle (*Liodes cinnamomea*) and truffle flies (such as *Suillia fuscicornis* and *Suillia gigantea*) are attracted by the odour of the fungi and deposit their eggs onto or into the fruiting bodies. Bacteria can enter at the same time, which can quickly cause the truffles to rot and give off a disgusting smell. Truffles can also turn mouldy, particularly during storage, or rot in wet ground. Bacteria can also turn the inside of Italian white truffles brown. Truffles that freeze in the ground and then thaw out will also rot in the ground, and this can be the cause of major losses in some years.

Slugs have been troublesome in a New Zealand truffière in years when the soil was wet. The slugs burrow into the truffles and eat the tissue inside. The control measures have been to place slug baits throughout the truffière; to deep rip the soil to drain and aerate the soil, albeit at the expense of considerable damage to the root systems; and to remove trees to ensure good air circulation and help keep the soil drier. Rats, mice, and moles can all cause problems as well, not just because they eat truffles but also because

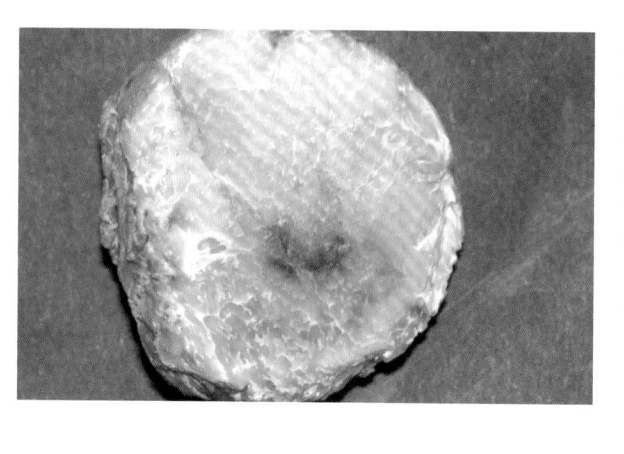

A variety of diseases can affect truffles. This fungal disease causes the insides of Italian white truffles (*Tuber magnatum*) to turn brown.
A. Zambonelli

their truffle-scented droppings can confuse the truffle dogs and pigs. French truffle growers suggest that fallen hazelnuts should be removed to avoid attracting rodents.

Pruning

Pierre Sourzat suggests that, to limit transplant shock, newly planted trees initially should be cut back to 5 or 10 cm above the ground soon after planting and before bud break has occurred. In most New Zealand truffières, however, transplants can be as much as 1 m high, yet survival and establishment rates were good. This success was, at least in part, a reflection of careful and frequent watering during the first few months after planting.

Whether further pruning is carried out after the shape of trees has been corrected depends on the climate and the species of host tree and truffle and is the subject of intense discussion. Burgundy and Italian white truffles fruit well under a closed canopy, so pruning is not particularly important. In Périgord black truffières, however, particularly those in cool areas, removal of the lower branches and suckers that shade the ground is probably important. Oaks generally have a strong apical dominance and need little pruning, except when they have been affected by trace element deficiencies, in which case the loss of the apical bud will require considerable attention to stop trees turning into a shape reminiscent of a weeping elm. In European truffières, oaks occasionally will be polled and heavily pruned. Trees that sucker, such as the hazelnuts, may also need considerable attention to remove the suckers with either pruning shears or glufosinate-ammonium. This is best carried out in late winter or early spring and before bud break has occurred.

When growing the Périgord black truffle, the aim is to produce trees preferably with a single trunk, 0.5 to 1.0 m high, and with sparse, well-spread foliage in the shape of an ice cream cone. This shape is chosen so that the oblique rays of sun warm the soil in the rooting zone in spring and autumn and early and late in the day in summer, while the foliage protects the soil from overheating during the middle of the day in summer. The principal effect of this is to raise soil temperatures in spring to ensure

Loss of the apical bud either because of physical damage and/or a trace element deficiency can lead to problems with tree shape that must be corrected by pruning. I. R. Hall

an early resumption of mycorrhizal activity, while preventing excessive loss of soil moisture during the summer months. Europeans have mixed views on whether pruning should be continued once production has started. In those New Zealand truffières with a high planting density and high growth rate, such as Alan and Lynley Halls' truffière at Gisborne, New Zealand, there is no question that continued pruning was absolutely essential to counteract the decline in soil temperatures, increased soil moisture, and reduced aeration associated with the development of a canopy—a situation that under other circumstances could also have favoured the growth of root pathogens such as *Armillaria*.

Thinning and the Rejuvenation of Truffières

In European mixed hazelnut and oak truffières, hazelnuts produce until they are about 25 years old, when it is considered necessary to remove the trees to leave only oaks. Once the canopies of the remaining oaks begin

A high planting density in an experimental bianchetto (*Tuber borchii*) truffière at Marina di Ravenna led to early canopy closure followed by the replacement of the truffle by contaminating fungi and a loss of production. I. R. Hall

to impinge on one another, it is necessary to remove some of them as well, until eventually there are perhaps as few as 50 trees per hectare. While thinning the oaks, the remaining trees are often given a very hard pruning by removing the slanting and horizontal branches and topping the leader. In this way, truffières can be made to continue producing for upwards of 50 years. However, because hazelnut can continue to be coppiced for several hundred years, there seems no reason why these trees cannot also be managed rather than removed.

In Alan and Lynley Halls' Périgord black truffière at Gisborne, New Zealand, the dramatic fall in production that accompanied canopy closure was only reversed when the original planting density was reduced from 800 trees per hectare to half this. Removing the trees allowed the soil conditions to recover to more like they were in the first decade of the life of the truffière. The answer to whether planting 400 trees per hectare instead of 800 would have resulted in an equally productive truffière will have to wait for the results from later plantings.

Maintenance of Truffières

In 1990 Alessandra Zambonelli and colleagues established a field experiment at Marina di Ravenna, Italy, to study the interaction of bianchetto and competing fungi. A very high planting density (2500 trees per hectare) was used. Bianchetto production began after only four years but decreased and moved to the border of the truffière as the canopy cover approached 100 per cent. Production recovered somewhat after thinning, but later collapsed again. These results might suggest that *Tuber borchii* is an early successional stage ectomycorrhizal fungus, which tends to occur in soils in the first phase in the development of a forest ecosystem. However, later field experiments and commercial cultivation of bianchetto with lower planting densities did not show similar falls in production. Consequently, to maximize bianchetto production it seems best to use a low planting density, thin the trees, and maintain a reduced level of undergrowth to arrest the replacement of bianchetto by late-stage ectomycorrhizal fungi.

SEVEN

The Rewards

NESTLED IN THE FOOTHILLS of the Apennines, Acqualagna and the neighbouring small towns and villages are surrounded by hills, mountains, streams, and valley bottoms with a wide range of habitats where truffles thrive. As a consequence, this is one of the few areas in the world where the tables of restaurants like La Ginestra in Passo del Furlo are graced with truffles almost year round. The Italian white truffle is available from October to perhaps the middle of January, Périgord black truffles from late November until the end of February, in a good year bianchetto between January and late March, and summer and Burgundy truffles from May through to the end of December (Appendix 13). In most truffle-producing areas of the world, however, just one or two species of truffle are harvested and the start of the season is keenly awaited. But any impatience must be curbed. Truffles harvested too early may not have had time to mature fully and will have little or no taste or culinary value. The timing of the truffle harvest is enshrined in national and local laws to ensure that the best-quality truffle is available for the market and not wasted by overenthusiastic harvesters keen to sell the first truffles, whether mature or not.

Le Cavage: **The Harvest**

Traditionally, when and how to collect truffles have been jealously guarded secrets, and knowledge of where to find them has been passed down from father to son, generation after generation. Truffle hunting is usually a solitary early- to mid-morning business, conducted discretely when the air is crisp and clear and the distinctive smell of a ripe truffle hanging in the air is not camouflaged either by the lingering scents of nocturnal animals or the odours elicited by daytime warmth. The truffle hunter moves slowly across stony, barren-looking ground beneath the skeletal canopy of winter woodland or plantation, in silence, senses honed—for finding the truffle in its subterranean habitat is a delicate and always uncertain enterprise.

An experienced truffle hunter can often deduce the position of an expanding truffle in late summer and early autumn from a mound in the soil and telltale cracks radiating from the centre. Because the cracks will eventually fill when it rains, the truffle hunter may mark the mound with a snail shell or a pile of stones or by planting a few seeds of barley nearby, which sprout to become botanical pointers. The hunter will then return months later to unearth the mature truffle.

In Italy people once dug for truffles just like digging for potatoes, but

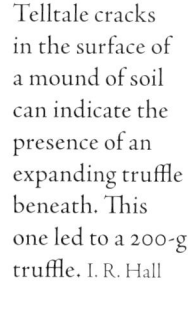

Telltale cracks in the surface of a mound of soil can indicate the presence of an expanding truffle beneath. This one led to a 200-g truffle. I. R. Hall

After a truffle dog has marked where a truffle might be found, the owner may sniff the soil before digging to confirm the presence or maturity of the truffle.
A. Zambonelli

this practice has long since ceased, as it is now known to have a detrimental effect on the mycorrhizas and can severely depress yields in subsequent years. Similarly, small adzes are used to unearth truffles in China, with the result that both mature and immature truffles with almost no aroma are harvested. Regrettably, because a Chinese harvester can earn more in a day collecting truffles than he might earn in a month at his regular job, the damage to forests caused by digging for truffles seems unlikely to stop in China. In North America and North Africa, it is also the norm to rake for truffles in likely areas, such as where rodents have been digging. Unripe truffles are invariably collected along with mature and overripe truffles. The truffles are then sold without grading, which has led to native North American truffles being branded inferior. This is not necessarily true, however, as fully mature Oregon white truffles have an excellent flavour that some argue is every bit as good as the Italian white truffle.

The Rewards

In Morocco desert truffles are harvested from the soft sand using a small rake.
A. Zambonelli

Harvesters of desert truffles look for them after spring rain, with surface hollows being a common habitat. Small mounds in the ground and cracks in the soil surface that look like crow's feet signify a swollen truffle beneath. The Kalahari Bushmen look for places where mongoose or bat-eared fox have been digging, while in parts of North Africa a pointed rod is pushed into the ground to detect the presence of the truffles.

Some truffle hunters have a keen sense of smell and can detect the buried treasure just by sniffing the ground. But truffle hunters have long relied on the assistance of animals and their more acute sense of smell to detect the giveaway aroma of a ripe truffle. Probably the longest practised method of finding truffles, one still favoured by many, is *avec la mouche* (with the fly). The larvae of certain flies feed on truffles, the most common being *Suillia fuscicornis*, *Suillia gigantea*, and *Suillia hispanica*. When truffles are ripe, swarms of these flies can be seen sitting on the vegetation or hovering over the area where the truffles are growing.

Some people like Juliet Fowler are able to find ripe truffles just by sniffing the ground. This was the only way some of the first New Zealand truffles could be found. Gerard Chevalier and Chantal Dupré (rear) could hardly contain their mirth. I. R. Hall

Hunting truffles *avec la mouche* involves walking through an area where it is expected that truffles can be found, gently disturbing the vegetation ahead with a long stick, and then looking for the truffle flies that spring up from where a ripe truffle is buried. While this might seem a bit hit-and-miss, looking only under the right types of trees, in brûlés, and walking towards the sun to avoid casting a shadow—so the flies can be seen sparkling in the sunlight when they rise—can considerably improve the chances of finding truffles. A vivid description of hunting truffles *avec la mouche* can be found in Gustaf Sobin's novel of obsession, *The Fly-Truffler* (2000):

> Following the edge of the oak woods, he'd tap the undergrowth as he came with a curious little branch he'd whittled for the very occasion. The branch itself had been stripped of everything but a narrow wedge

Hunting truffles *avec la mouche* (with the fly) involves walking through a truffière, disturbing the foliage ahead with a long stick, and watching for where the truffle flies spring up. A. Zambonelli

The tiny truffle fly (*Suillia* sp.) can cause problems well beyond its size but can also be used to find truffles. P. Sourzat

of pine needles at its extremity. Tapping, patting the undergrowth, he could easily have been mistaken for a blind man. . . . Cabassac held the sun fixed like a compass point directly before him. That way, he knew, he wouldn't be casting a shadow over the brittle winter grasses: the very haunt of those tiny, straw like insects. Wouldn't be disturbing those diptera until the very last second. Because it was their sudden, spasmodic flight that betrayed their secret. They'd spring

rather than fly, revealing as they did the exact point . . . in the heavily scented earth directly beneath, that they'd lay their eggs. . . . that—by a miracle of pure symbiosis—one of those black, odoriferous tubers could be found.

Hunting with the fly likely dates from before recorded history. The use of goats has been reported in Sardinia, and bear cubs were once used to find *Choiromyces* near the convent of Sergiensky, Russia. The classic image of truffle hunting, however, is a pig on a leash in the French countryside. This method probably dates back to Roman times, although Platina (d. 1481), historian of the Popes, was the first to record their use. English naturalist John Ray, in a record of his travels between 1663 and 1665, also referred to the use of pigs for truffle hunting in Italy.

People have long noted that pigs, more particularly sows, have a natural fondness for truffles. Scientific research has shown this is at least partly because the fungi produce a steroid identical to a pheromone produced by boars during premating behaviour. The hormone identified, androstenol, is also secreted by humans, but in much lower concentrations than in male pigs, a fact that could be cited to support the aphrodisiac claims made for truffles.

Despite a natural fondness for truffle, some pigs are better than others and most require some training. According to Sourzat (1981), a three-month-old pig can be trained in two or three weeks. If a pig is to be purchased, a truffle is hidden in the hand of the buyer and the first pig who is tempted by the smell and approaches is chosen. The pig is taught to walk on a lead properly, for at least an hour a day, during the week before commencing the training in earnest. If the pig is more interested in searching for worms in the truffle bed, he is punished by a light tap with a stick. The pig seeks the truffle more for himself than to please his master. The training is based on a principle similar to that used for truffle dogs, even if it all starts with the taste of the truffle. Once accustomed to its taste, the object of the pig's desire is replaced by another delicacy. By tugging hard on the lead or cord, the pig will abandon the truffle in favour of a few kernels of corn or a piece of cooked potato, and the master then harvests the truffle.

ABOVE While many people associate hunting truffles with pigs, today they have been largely replaced with trained dogs. Courtesy of the Hall Family Trust

LEFT Hunting with a pig can be hazardous when the animal prefers the prized truffle to a piece of turnip. J. Parker and J. Parker

Though pigs are generally considered superior truffle hunters, dogs have gradually become more popular, largely because they are easier to handle. Retrieving a truffle from a sow driven by desire has its risks, and the loss of a finger might be considered an occupational hazard for the truffle hunter. In addition, dogs have more energy than pigs, fit better on the back seat of a car, and are less conspicuous—an important consideration for the truffle poacher tempted by the promise of high returns. Taking a pig for a walk is a bit of a giveaway.

Unlike pigs, dogs are not naturally interested in truffles but can be trained to indicate where they scent a truffle. No breed of dog instinctively looks for truffles, but there are a host of excellent truffle dogs that fall into a category some Italians call *bastardino*, whereas others use the more polite expression *meticcio*. Hunting dogs, however, are very difficult to train as they are more inclined to scent game than truffles. Many truffle hunters have their favourite breed of dogs. The Lagotto Romagnolo has become very popular in Italy, as its thick, shaggy coat provides protection from the cold in the mountains while truffle hunting during winter and its strong constitution will keep it going for hours. Pierre Sourzat swears by Labrador retrievers, while Diane Baker has a passion for Dobermans. A set of jaws running from shoulder to shoulder also gives the owner of a Doberman and a truffière another distinct advantage over one who favours the Chihuahua.

The method for training dogs, employing an association between finding a truffle and being rewarded with food morsels, appears to have originated in Italy and continues to be used today. Ramsbottom (1953) wrote, "At the beginning of the 18th century there are several records of dogs having been imported thence to Germany, where for a period, the possession of a 'Truffelhund' seems to have been regarded almost as a badge of nobility." He went on to say, "Frederic William of Prussia imported trained dogs from Lombardy, the chief breed of which was a kind of poodle," and speculates that truffle dogs were probably introduced to France at about the same time. Reference to poodles as truffle dogs was also made in *Der Deutsche Pudel* (1907):

The Lagotto Romagnolo is a dog breed favoured by the truffle hunter because its thick, warm coat enables it to continue working in cold weather. Dino Grammatico holds a long-handled trowel used to find Italian white truffles, which may be up to 0.5 m belowground. A. Zambonelli

Another use for the Poodle, because of his excellent nose, is that of truffle-dog. As early as 1746, Döbel in his Jägerpractica says that in this he is superior to all other breeds, and is nowadays used for this purpose in Italy, Spain, and France and in those areas of Germany where truffles exist. . . .

Training should begin in the summer. . . . Initially, one practices retrieving at home by sewing strong-smelling truffle into a small leather pouch and hiding this in various parts of the room and making the dog find and retrieve it. Once he does so willingly and without "mouthing," it is then hidden in shrubs or under moss or leaves. . . . Later the pouch is hidden under deeper layers of moss or leaves. Once the dog finds it quickly and surely, one takes him to parts of the forest where truffles are known to grow, and makes him find and retrieve under ever more trying conditions. During training, one should talk to the dog as little as possible, not using any words other

than those commands chosen for the truffle hunt. When the dog has found the truffles, he is to be praised and patted and made to sit, then rewarded with a "treat."

Not until the middle of September after it has rained does one take off the leash. . . . [In] areas in which truffles are suspected, there one hides a fresh truffle (not in a pouch) on a number of occasions under two inches of earth and leaves. If the dog scents the truffle and starts to scratch to recover it, one should quickly go to the location and recover the fruit oneself, showing it to him and putting it into one's pocket. From now on the dog should not be allowed to retrieve, but merely indicate the location. When he does so, he should be made to sit, praised, and rewarded.

Once the dog discovers the truffles even underground, take him into a truffle region and cause him to quarter upwind, since this will cause the scent to be found faster and more easily. An experienced and eager dog can scent truffles at 20 paces, and 6 inches below ground. . . .

After returning from the hunt, the dog should be fed his normal food, never before the hunt, when he should not get anything but fresh water. During the hunt he should be given treats only when he has actually found truffles. If he finds nothing, which can happen in dry and hot weather, he should not be given anything other than his food upon returning home.

Rawdon B. Lee's *A History and Description of the Modern Dogs of Great Britain and Ireland* (1894) also mentioned the training of poodles for truffle hunting, with a more recent, less-breed-specific treatment of the subject being detailed by Rocchia (1992). Another method favoured by New Zealanders who train dogs to find drugs or illegally imported fruit uses play as a reward for finding the quarry rather than food—an important consideration for someone confident of finding huge quantities of truffles and with the nightmare of a dog already full of treats, lying cast on the ground, with half the truffière still to search.

The cost of a trained dog, including the purchase price and training, can run into several thousand euros. Like the quarry for which they are trained, truffle dogs are sought after by fair means or foul. According to a report in the *Guardian* from January 2003, Michel Tournayre heard a car drive up to his truffière in the hills above Uzès, a town in the Gard region of France. He came out in time to see a vehicle drive off with his dog, who was born into a line of truffle-hunting dogs and was worth about US$5000. Tournayre said the gang was unlikely to have sold her, instead claiming she was being used to steal truffles from people like himself. His was one of at least eight truffle dogs to have disappeared over the previous year in southern France.

Problems associated with training and keeping animals for truffle hunting have stimulated the search for an "electronic nose," such as that developed in the mid-1990s in a joint research project between the University of Toulouse and the University of Manchester. While not necessarily designed as a replacement for dogs or even pigs, it was believed the detector would have the advantage of not tiring as the hunt for truffles wore on. Because the detector can distinguish between different volatile compounds and use this information to identify only truffles mature enough to have the best flavour, it should also be more accurate. Although the detector did prove more accurate in trials, it was much slower than a dog.

While such developments are based on sound science, there are also people in Italy and France who are convinced that truffle activity under the ground can be located or identified with the use of divining rods or pendulums. This notion has been taken seriously enough by a section of the truffle industry that articles have been published on the subject.

Regulations and Reality

It has been said that for every truffle harvested legally, another is poached. Laws enacted to protect the industry may be backed up in truffle-growing areas by signs warning of the consequences of illegal harvesting. Some of the signs promise retribution well beyond the bounds of the law, as illus-

trated by the following example quoted by Peter Mayle (1999, p. 5), "*Tout contrevenant sera abattu, les survivants poursuivis*" (Trespassers will be shot, survivors prosecuted).

The delicacy and danger of the situation was brought home to Ian Hall on a visit to France soon after beginning his quest to produce truffles in New Zealand. Having boldly walked into a truffière to take photographs, he was warned by his white-faced guide, "Do not do that again. If you do, you will be probably be shot having been mistaken for a pigeon." In addition to signs, large dogs are kept as deterrents. The prize for innovation, however, must go to a flamboyant Frenchman who claimed he protected his prized truffière with land mines. And who was going to test his honesty?

While many stories concerning truffles are anecdotal, the threat of poaching is real enough. In January 2003 a story appeared in the newspapers reporting that truffle producers in the Var region of southern France had armed themselves with guns to protect their oak plantations, where it was claimed organized gangs of thieves had been digging up truffles by night and selling them at the local markets. The situation reached a climax on Christmas Day, when there was an exchange of gunfire between the producers and one of the gangs. Although it was too dark for either side to see clearly and no one was injured, the incident prompted the head of the local association of truffle producers, Francis Gillet, to call for calm.

A French sign warning that a truffière is protected, that entry is forbidden, and the relevant legislation
I. R. Hall

TRUFFIERE PROTEGEE PAR LA LOI

Avec l'assistance des

SERVICES du MINISTÈRE de L'AGRICULTURE

DEFENSE DE PENETRER

sous peine de sanctions correctionnelles graves

Art. 388 et 444 du code pénal

Nevertheless, the producers said they would continue to mount an armed guard around their oak trees until the end of the truffle season.

Suffice to say, more reasoned and formalized rules and regulations have been enacted in Europe. In Italy there is a national law that covers the most important rules for harvesting, cultivation, conservation, and marketing of truffles. For example, immature truffles are protected, dogs must be used to locate truffles, a special small trowel must be used to excavate the truffles, and excavations are to be covered immediately after a truffle is harvested to avoid damage to the mycorrhizas. In addition to national laws, there are also local laws that define how the national law will be applied in the regions, such as the official harvesting times for each species of truffle (Appendix 13), the cost of permits, and the regulations regarding access to controlled and cultivated truffières. For example, all truffle collectors have open access to the extensive oak forests and woodlands and other lands not under cultivation but must pass an examination and be licensed before they can legally gather truffles. Truffles under managed production are the property of the landowner and are subject to separate regulation. The landowner must first ask permission from the local government to close a natural truffière, however, and the regulations controlling this differ somewhat from region to region. Similarly, in most parts of Italy truffle hunting is allowed only during the day, but in Liguria, Lombardy, and Piedmont forays into forests are also allowed at night. There are also strict regulations governing quality standards for fresh and preserved truffles. Details of these laws are listed on several Web sites, and the United Nations has recommendations for the marketing and quality control of fresh truffles.

A Time-Honoured Ritual

During *le cavage*, the rabassiers of southern France and the tartufai of Italy and buyers meet daily in the cafés and bistros of towns and villages, where discussions, lively and furtive, may range from immediate concerns with yields and prices to wider issues facing the industry, including its continual decline in Europe, the success or otherwise of cultivation, and, in recent years, the threat posed by cheap imports from Asia.

For such harvesters, a bountiful truffle crop is entirely dependent upon the right growing and weather conditions, and a bumper harvest means big money. Theirs is the traditional face of an industry that has gone global, where a worldwide clamour for the best of the small offering that Europe can produce has pushed prices to extraordinarily high levels, stimulated research and investment, and fuelled both the blunt and innovative efforts of illegal elements that have long shadowed the truffle world. Regrettably, despite the growth in rules and regulations, no section of the truffle industry is immune from illegality, whether it be the early-morning plantation raids and dog thefts, marketing and investment fraud, or restaurant misrepresentation.

Advances in truffle science and knowledge of the growth process have, as yet, had little impact on a European decline. The ravages of war, on both men and countryside, urbanization, the clearing and/or degradation of woodland, pollution, the adoption of large-scale and intensive farming practices, and adverse climatic conditions have all contributed to this decline.

Efforts to cultivate *Tuber magnatum* have not yet been completely successful. While it can be said the codes of species such as the Périgord black truffle and Burgundy truffle have been cracked, cultivation remains precarious and labour intensive. Despite the possibility of high rewards, the time involved and potential for failure can be off-putting considerations for would-be trufficulteurs in Europe.

Failure to Produce

Despite historical factors and the vagaries of climate, the failure to arrest the decline in truffle production still leaves cause for wonder, considering scientific advances; ongoing cooperative efforts by Italy, France, and Spain to turn the situation in Europe around; and the worldwide planting of thousands of truffle-infected tree seedlings—France alone plants an average of 400,000 mycorrhized seedlings each year. A closer look at just why people plant truffle-infected tree seedlings suggests some underlying reasons. In some cases the production of truffles may be a secondary motiva-

tion, a bonus, and in others the initial enthusiasm to do so may wane when ongoing investment costs are weighed against the uncertainty of future rewards and be distorted by other considerations. For instance, under European Union government policy, the planting of trees is rewarded with cash grants, thus money is made whether truffles are produced or not. In New Zealand and Australia plantings have been used to justify the subdivision of farmland into high-priced residential units where land-use regulations and zoning laws designed to ensure the productive use of land would otherwise prevent such an activity. In the United States the attraction of tax breaks may be a major consideration.

Such scenarios suggest many plantations may not be looked after to the extent needed to ensure fungal development, even if climatic and soil conditions are conducive. Add to that the possibility that some of the many tree seedlings planted by would-be truffle producers have been inexpertly, inappropriately, and/or fraudulently inoculated (Chapter 2) or have been planted in or near ground where competing fungi are present, and what on the surface looks like a huge investment in the future of the truffle industry may in reality be considerably diluted.

Perhaps the worst case of outright inoculation fraud that we know of occurred in Italy some years ago. A company took out full-page newspaper advertisements announcing truffle-infected plants for sale. Although the price of the plants was very high, the production of truffles was guaranteed in a few years and the company promised to buy the truffles at a high price. After a year or so the company inspected the roots of plants in investors' truffières and found that many plants were not infected with truffle. The company then offered to provide a remedial treatment for "only" €25 per plant. This consisted of putting a yellow liquid around the bases of the uninfected trees—a treatment that subsequently was found did not even contain truffle spores. When the time came for the dogs to go out to find the first truffles, all traces of the company had vanished.

Many cases of dishonesty are more innocuous than the above examples, but the ultimate result may be just the same. For example, truffle plants are still widely marketed as "inoculated" rather than "guaranteed infected with a particular species of truffle with x per cent of the root tips converted to

mycorrhizas." Similarly, nurserymen would never reveal to their clients that the plants they were selling may have been inoculated with a serious mycorrhizal competitor such as *Tuber indicum* or that they were contaminated with one of the many other nontruffle ectomycorrhizal contaminants.

Cultivation Outside of Europe

The first Périgord black truffles to be harvested outside of Europe came from a northern Californian truffière owned by Bruce Hatch, which began producing in 1991, nine and a half years after it was planted with trees obtained from Agri-Truffe in France. A second truffière in North Carolina, also established with hazelnuts from Agri-Truffe, began producing in 1993, 12 years after planting and only after the owner had applied large quantities of limestone and hydrated lime to raise the soil pH. In contrast, a truffière near Gisborne, New Zealand, produced its first truffles in 1993, only five years after planting, perhaps because of the very rapid growth of the trees, adequate soil moisture, or the absence of competing fungi.

The possibility of high returns has fuelled interest in growing truffles worldwide, but the development of truffle cultivation in nontraditional areas and countries—the United States, New Zealand, and Australia—has yet to make any significant impact on the shortfall in supply and the ever-growing demand. For example, before the mid-1990s, New Zealand did not have a truffle industry at all and demand for truffles there was almost nonexistent. Subsequently, Périgord black truffle production in New Zealand and Australia has increased yearly, but so too has local demand, fuelled by publicity and greater culinary awareness, to such an extent that domestic production cannot keep pace and very little of the harvest is sold on export markets.

More than 300, mostly Périgord black truffle, truffières are currently established outside of Europe, with more than 100 in New Zealand, a similar number in Australia, approximately 120 mostly small truffières in the United States, and a number in Argentina, Chile, Israel, and South Africa. Most of these truffières are too young to produce or too small to produce large quantities, but some spectacular results have been obtained.

Yields

Reliable information on truffle yields and production is very difficult to obtain, primarily because of underreporting of harvests, under-the-table marketing practices, and a lack of government records. Truffle production, as with other mushrooms, also varies dramatically from year to year depending on weather conditions. Thus, even with reliable data, establishing average productivity values would require many years, while measuring trends would likely require decades. However, it is indisputable that there has been a catastrophic decline in Périgord black truffle production over the past 100 years, from more than 1000 tonnes at the start of the 20th century to only 40 to 150 tonnes now. In years when climatic conditions are unfavourable, this level of production may be cut by half or more, as illustrated by the 1998–1999 season and even more graphically the 2003–2004 season, when the harvest was considered to have been the worst in living memory (Appendices 4, 5). Similarly, the collective yields in France and Italy in 2005–2006 were low due to a very dry autumn in France followed by very low soil temperatures, which froze and destroyed those truffles that did form, coupled with very wet winter conditions in parts of Italy, which caused the truffles to rot in the ground.

The relative proportions of truffles harvested from natural areas and from plantations are similarly difficult to estimate, with experts undecided on which is currently the larger source. Under ideal conditions production in artificial truffières can begin after 3 years, but more often production does not begin until 7 to 10 years after planting, with yields then gradually increasing for the following 10 to 20 years. A smooth increase in production is unlikely because of bad seasons, largely the result of inadequate summer or autumn rains. The long-term pattern of yield will also be affected by the number of truffle-infected trees per hectare, the host species that have been planted, whether an irrigation system has been installed, how good the truffle dog and its handler are, and the like.

In Europe many truffières are not irrigated, and techniques are used that conserve the effort required to maintain them at the cost of lower yields. Average annual yields of 15 to 20 kg/ha are considered good results (Chap-

ter 4). Franklin Garland has reported harvests of approximately 50 kg/ha in North Carolina, which are comparable to production in several New Zealand truffières. In comparison, a natural truffle-growing forest would not be expected to produce more than 10 kg/ha.

In the booklet *Truffle: The Black Diamond* (1980), François Picart stated, "production after the 7th year should range from 1/8th to 4 lbs per tree." Assuming that 400 trees were planted per hectare and 80 per cent of these produced, these yields work out to between 18 and 580 kg/ha. This booklet was written in the late 1970s, when there was relatively little published information available on the likely yields of artificial truffières. Consequently, the predictions made in it were probably convenient, optimistic extrapolations of relatively high yields found under highly productive individual trees in Europe. There are, however, instances of exceptionally high yields greater than 100 kg/ha in commercial truffières in Europe and New Zealand. For example, Gerard Chevalier (1998) listed the best results he had observed in Burgundy and Périgord: the host trees were *Corylus avellana* and *Quercus pubescens* ranging in age from 11 to 14 years, and yields ranged from 100 to 150 kg/ha. The returns from such truffières are also exceptional given that at the time annual harvests of only 8 to 10 kg/ha were required to recoup maintenance and investment costs in France.

Prices

Truffles are renowned, in part, because of their high prices. Prices vary from season to season in response to supply and demand and within seasons depending on availability and proximity to Christmas, when they are generally the highest. Prices also vary by source locality and country. Black truffles from Périgord command the highest prices, while *Tuber melanosporum* from other parts of France and particularly from Spain and Italy tend to have somewhat lower prices. Different species also are priced differently, with the Italian white truffle by far the most expensive, Périgord black truffle a distant second, and several species of black truffle from Asia at the lower end of the scale.

In 1992 approximately 32,000 kg of Oregon white truffles were harvested

in Oregon and Washington and sold for an average price of US$85 per kilogram (US$32 per pound), about one-third of current prices. In comparison the desert truffles command a paltry sum, with retail prices in Spain fluctuating between €12 per kilogram when the truffles are abundant and up to €30 per kilogram when there is a shortage, although in Moroccan marketplaces prices during the middle of the season can be as low as €3 per kilogram.

The price for the Périgord black truffle in Europe increased steadily during the second half of the 20th century, and in 2003–2004 there was a sharp rise due to a climatically triggered fall in supply. The yield for Périgord black truffles in that season fell to around 20 per cent of what would be expected in a normal year, and wholesale prices surged to more than $1250 per kilogram. The disaster of that season was attributed to a very hot and dry summer. Ground temperatures of up to 40°C desiccated the trees and the mycorrhizas, adding weight to the old French proverb "water in July, truffles at Christmas." The same trends were observed for the Italian white truffle, the harvest of which was equally hit by the drought of 2003. The

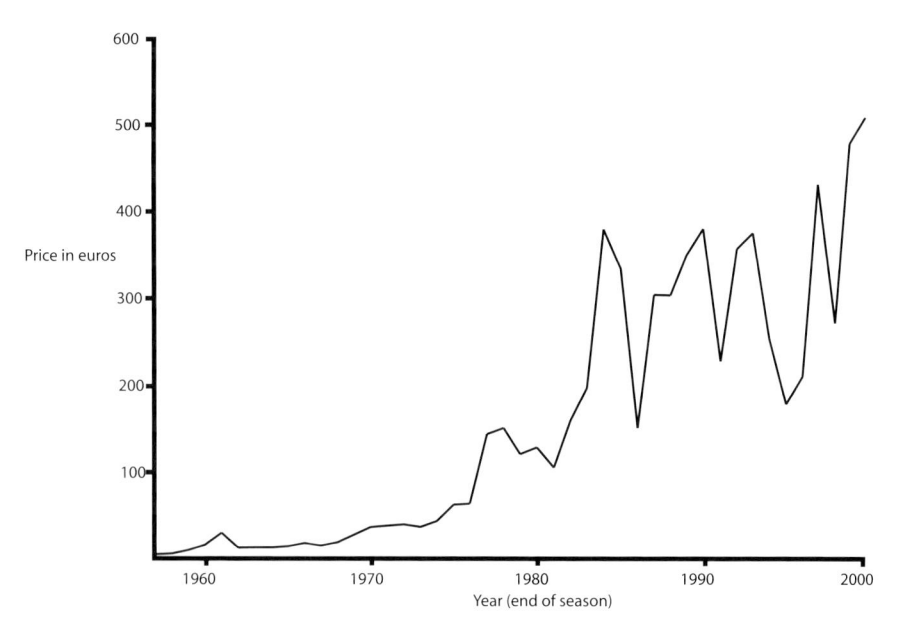

The price of the Périgord black truffle (*Tuber melanosporum*) rose steadily over the second half of the 20th century. Data from Sourzat (2002)

wholesale price for Italian white truffles in 2003–2004 ranged from €2000 per kilogram for small specimens to €3500 per kilogram for large ones. At the same time the Burgundy truffle sold for €500 per kilogram and the bianchetto truffle about €350 per kilogram. On the retail market, at specialized boutiques such as La Maison de la Truffe in Paris, the price for a kilogram of fresh Périgord black truffles was around three times as high. By Christmas 2005 the price of the Périgord black truffle at Harrod's of London had hit £2000 per kilogram (€3000 per kilogram) and for the Italian white truffle double this.

Prices for Périgord black truffles grown in the Southern Hemisphere tend to be higher than those in-season in the Northern Hemisphere, reflecting both the relatively low level of supply at a time when no fresh European truffles are available and the premium that buyers are prepared to pay. In 2006, for example, grade 1 New Zealand truffles sold at the farm gate for US$2500 per kilogram, compared with US$1550 to US$1900 per kilogram at French, Italian, and Spanish wholesalers six months before.

Such prices may appear high enough, but each year newspapers carry reports of even more breathtaking prices paid at specialist auctions, usually in Alba. Admittedly these prices, rather than being market driven, are determined by the pursuit of publicity and promotion and a desire to contribute to charitable causes—such as flood victims in Piedmont in 2000 and earthquake victims in southern Italy in 2002—but they also send a potent message of exclusivity, demand, and potential profits.

In November 2000 a British newspaper reported white truffles had become more expensive than gold as a result of an ever-dwindling supply and soaring demand. At an auction near Alba, where buyers in Tokyo bid via a television linkup, restaurateurs fought for prized examples of the Italian white truffle in which a record £5000 (approximately €7400) was paid for a 497-g truffle, equating to £280 an ounce (28.35 g). Gold at the time was being quoted at £185 a troy ounce (31.1035 g). In November 2002 a Los Angeles restaurateur paid US$35,000 (approximately €27,000) for an Italian white truffle weighing just over 1 kg at a similar auction in Alba. Late in 2004 the record was broken again when a consortium paid US$53,000 for an 850-g Italian white truffle from San Miniato in Tuscany. But the

story does not finish there—the truffle went on show in a Knightsbridge restaurant and deteriorated to a point where it couldn't be used. The truffle was sent back to Tuscany, where it was wrapped in a blue cloth and, accompanied by Tuscan guardsmen in medieval costume, it was solemnly interred under trees in the grounds of the Castello di Cafaggiolo, where it was hoped it would produce mycorrhizas and more giant Italian white truffles. The 2006 season saw the top price reach even giddier heights, when a 1.5-kg Italian white truffle sold for €125,000 (US$160,000; £83,400) to a Hong Kong buyer at a charity auction in Italy.

Such heady prices are a long way from 1894, when fresh truffles on the Cahors market sold for an average price of 12 s / 6 d per kilogram (£0.63 today). Twenty years later the 1913–1914 edition of the Harrod's catalogue included several entries for truffles: one-quarter pint of the "finest French extra choice in bottles" sold for 1 s / 6½ d, a pint of "extra choice in glass bottles" ranged from 5 s / 3 d to 6 s / 3 d, a pint of "finest truffles in tins" cost 5 s / 3 d, and a pint of "truffle peelings" sold for 1 s / 7 d. If we assume that the value of money in the United Kingdom has depreciated 60-fold since 1913, then a quarter-pint bottle of truffles (about 100 g of truffle) would now cost about £4.50 (about US$8.00), which is a small fraction of the current price of the same product in Harrod's today.

Marketing by Fair Means or Foul

The perishable nature of truffles, as with all mushrooms, has been and remains a determining factor in their successful marketing and distribution. In ancient times, the Romans imported the terfez or desert truffle from North Africa and the eastern Mediterranean in sealed jars filled with sand. Today, European, Chinese, and New Zealand exporters use or are developing sophisticated chilled packaging technology to airfreight fresh truffles from the point of harvest to the point of consumption within two to three days. Companies, particularly those in Europe, bottle, can, freeze, vacuum seal, store using modified atmosphere technology, and process truffles into food products, such as foie gras, both to ensure their availability out of season and to exploit a diminishing resource to the greatest extent possible.

Despite the involvement of big business and the advances that have been made over the past 150 years or so, most truffles still come from individuals harvesting relatively small truffières, and a method of marketing has grown to suit. While some of the harvest is sold under the counter, local markets, with time-honoured rituals and careful weighing, are held weekly to cater to the many small lots of truffles offered for sale. To avoid the spectre of taxation, it is predominantly a discreet, cash-only business.

Lalbenque in southwestern France is the site of one of the more important Périgord black truffle markets. On one day of each week during the truffle harvest, the previous week's collection of truffles are displayed on trestle tables set up around the market square. With the collectors standing behind their produce and the representatives of the canning industries, the wholesalers (the men in the Mercedes), and private buyers moving from table to table, when there is much sniffing and animated haggling. Quantities, offered prices, and counter requests may be written on small slips of folded paper that are passed to and fro until a deal is reached. It may be regrettable, but some of the harvesters, reluctant to deal with anyone they do not know, have opted out of this quaint system in favour of supplying truffles on contract or selling directly to wholesalers, restaurants, or shops, perhaps accepting less than top market prices in exchange for the security of dealing behind closed doors.

Truffle hunters and merchants can earn their yearly wage during the short three- to four-month season in trading that can involve shady transactions, fungal substitutions, and other less-subtle deceptions. Slivers of metal or lead shot have been inserted in truffles or in the soil covering unbrushed truffles to make the baskets heavier. Old truffles or those riddled with insect larvae have been frozen to make them firmer. Old, dried-out truffles have been moistened and included with fresher truffles to temporarily take on their fragrance. The holes and dips that occur naturally in truffles have been filled with mud, and two small truffles have been stuck together with a toothpick and packed with mud to appear as one large truffle.

As white truffles from Piedmont and black truffles from Périgord command the best prices, the same species or similar may be brought in from other regions or countries and sold for a price higher than they would get

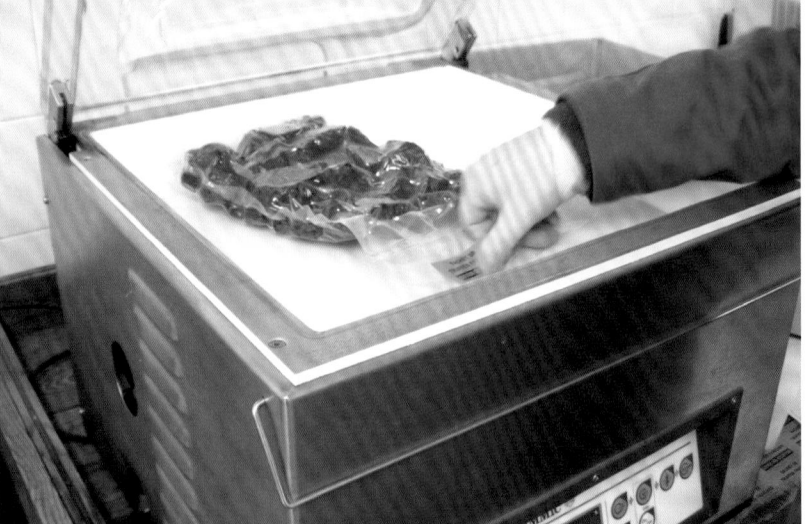

TOP Vacuum sealing truffles in plastic bags is a useful way of sending truffles by refrigerated transport. I. R. Hall, courtesy of the Pébeyre family

ABOVE LEFT The traditional way of preserving truffles was to bottle them. This handmade bottle dating from about 1820, complete with its original contents of three truffles, cork, wax seal, and label, turned up in a bottle collection in New Zealand a few years ago. The label has a portrait of Napoleon and reads: "Truffe du Périgord, Productions du Provence, Fruits & Confits, J. Dissard & Co. Sucrs, Londres & Carpentras, Napoleon Emporeur." I. R. Hall, courtesy of R. Roberts and C. Roberts

ABOVE RIGHT Truffles are often preserved by canning. Here the truffles are being checked to ensure they are free of adhering soil. P.-J. Pébeyre

TOP Animated haggling over the price and quality of truffles is part of the scene in the Lalbenque truffle market in southwestern France. I. R. Hall

BOTTOM Large quantities of wild Asiatic species of truffles are offered for sale in the Nanhua mushroom market, China. Y. Wang

where they were harvested. Similar species that may be passed off as the Italian white truffle include *Tuber borchii, T. dryophilum, T. maculatum, T. oligospermum*, and *Choiromyces meandriformis*. Species substituted for the Périgord black truffle include *Tuber brumale, T. aestivum, T. mesentericum, T. macrosporum*, and several Asiatic species, including *T. indicum*. With China the largest producer and exporter of Asiatic truffles, these less-valued species are often referred to collectively as "Chinese truffles." They look very similar to the Périgord black truffle but do not have its pungent aroma and culinary properties. Based on their external appearance, mature specimens of Asiatic species such as *T. indicum* and the Périgord black truffle are difficult to distinguish, and even when cut are easily confused. To distinguish them properly requires a high-powered microscope.

In the absence of microscopic assistance, and despite a huge difference in the intensity and quality of the aroma of the Périgord black and Asiatic truffles, a buyer may be fooled if the Asiatic truffles have been kept in a closed container, thus concentrating the aroma. Alternatively, the less-precious winter truffle or Asiatic truffles can be mixed with Périgord black truffles to confuse the buyer. As a rule of thumb, fresh black or white truffles sold outside the normal harvest times of *Tuber melanosporum* and *Tuber magnatum* invariably are other species, simply because truffles cannot be refrigerated for more than a few weeks, when their distinctive aroma begins to fade and the truffles themselves quickly turn into a putrid mass. Other truffles may also be treated with chemical additives that approximate the sought-after aroma of either the Périgord black truffle or Italian white truffle. In fact, while truffle volatiles can be captured in olive oil, the vast majority of white and black truffle oils have never seen a truffle and instead are made with the chemicals *bis*-methylthiomethane, methyl-2-butanol, methyl-*b*-2-butanol, methyl-2-propanol, acetaldehyde, butanone-2, ethanol, and/or anisol. Similarly, any food such as oil or cheese with a distinct, continuous trufflelike smell has likely had a flavour enhancer added and probably will contain hardly any of the real thing, if any at all.

In other instances, experience is the key. While some sections of the industry seem to have particular difficulty separating the Périgord black truffle from the winter truffle, generally this should not present too many

problems if the truffles are mature. For several decades European authorities seemed to turned a blind eye towards those inventive souls who tried to make a diminishing crop go further. For example, in 1993 France imported 20 tonnes of Asiatic truffles to extend its own meagre harvest of Périgord black truffles that year. This move was the thin end of a wedge that opened a chasm of concern in traditional truffle-producing areas, but it is a practice that continues unabated to this day.

In subsequent years and during times of scarcity due to harvest failure, producers have become concerned—with justification—that restaurants are secretly substituting inferior and far-cheaper Asiatic truffles to stuff veal sausages, stud terrines, and dress up the region's famous sauce Périgord. The threat of European markets and possibly truffières being invaded by Asiatic truffle species and the declining reputation of the industry finally prompted an official response. Revised legislation was introduced in France, while in Italy there was a strengthening of the resolve of those entrusted with implementing the existing Italian law. Now, in France Asiatic truffles must be washed before sale and labelled correctly, and it is illegal to import them into Italy. The nine indigenous species that Italy recognizes as suitable for sale are the same as those recognized by France (Appendix 13), and both countries employ inspectors or judges to identify and grade truffles for quality. To aid this inspection, Périgord black truffles are often notched to reveal internal structures that demonstrate that they are not one of the Asiatic species. In France the preparation of truffle dishes in restaurants using Asiatic truffles without declaring it is punishable by two years in jail, fines, or both.

Having regulations is one thing, but the observance or enforcement of them quite another, and cheats are incorrigibly inventive. By the late 1960s, French regulations already restricted the word *truffe* to species of *Tuber* and marketed black truffles were to consist only of Périgord black or winter truffle, with a small concession (2 per cent) made for the Burgundy truffle due to the difficulties of identifying closely related species without magnification. Italian law is a little more flexible, allowing 3–8 per cent of truffle in pâté or in food products containing small pieces of truffle to be a species other than that on the label. The use of summer truffles (*T. aestivum*) in

LEFT This truffle-washing machine is a modified mussel washer. I. R. Hall, courtesy of Tofani Tartufi

ABOVE It relies on copious amounts of water and a rotating brush at the base of the vessel. I. R. Hall, courtesy of Tofani Tartufi

A tray of immature Périgord black truffles (*Tuber melanosporum*). Some show the small slices cut from the truffles to reveal the immature, pale interiors. I. R. Hall, courtesy of the Pébeyre family

pâtés is also regulated in Italy, and artificial colouring of white truffles to make them appear black is illegal. However, in 1988 an analysis of 74 truffle-containing pâtés found that 69 contained either desert or Burgundy truffles, 3 contained the winter truffle, and only 2 contained Périgord black truffles. In 1990 a truffle-faking ring was reported staining inferior, light-coloured truffles with walnut juice and passing them off as Périgord black truffles. One report said that Périgord growers did not object to the practice as there had been no complaints from restaurants, and, with supply well short of demand, the increase in business by fair means or foul was justified.

Today, inexperienced buyers can still be lured by Web pages offering "black truffles" for US$50 per kilogram, even when the seller is unequivocal about the species they are selling and requires a minimum order of 40 kg. The buyers may think they are getting a bargain but will surely be disappointed by the aroma and taste of what, at that price, can only be one of the lesser Asiatic truffles. Similarly, the inexperienced might be led to believe that high-priced preserved Périgord black truffles are as good as fresh. They are not. The quality and intensity of the aroma and flavour of a preserved truffle does not come close to that of a freshly harvested truffle. Canneries have been known to take fresh truffles and cook them several times over. Each time, the truffle-flavoured cooking water is saved and used to flavour lower-quality truffles that may then be sold as black truffles.

To respond to the range of illegal practices associated with the truffle trade, France has the Services de la Repression des Fraudes (Fraud Suppression Agency) to better regulate the industry. At the request of the consumer protection department of the French Finance Ministry, the National Institute of Agronomic Research and Antonella Amicucci and colleagues in Italy have developed sophisticated molecular techniques for detecting Asiatic truffles in food products. French researchers have also developed electrophoretic and immunological techniques to detect additives such as egg albumin and lactalbumin, which could be used fraudulently in truffle preserves.

There is still an important market for preserved truffles and truffle products, but increasingly the global demand is for fresh produce. While truffles from the Istrian Peninsula in Croatia may be harvested in the morning, wrapped in that day's newspaper to prove their provenance, and flown

to buyers in London, not all markets are within such convenient proximity. Fortunately, the truffle is more robust than the button mushroom and can withstand from one to three weeks refrigeration at 2°c, although flavour and texture do deteriorate with time. Thus, it is important that as few delays as possible lapse between the time of collection and final use and that the truffles are refrigerated whenever possible. Consequently, Italian truffles exported to the United States are washed, packed, delivered by hand to the airport, met at the airport in New York, carried through customs, and then repacked and dispatched to the consumers, all within 48 hours. In 1989 the United States imported 5.4 tonnes of fresh or chilled truffles in this way, mainly from Italy and France. These truffles had a value of us$1.48 million, or about us$380 per kilogram, a far cry from current prices.

France, Italy, and Spain are the major exporters of Périgord black truffles, and Italy exports white truffles primarily to restaurants in the United States and other high-priced markets in Europe and Japan. However, the growth in truffle production in the Southern Hemisphere, so far primarily Australia and New Zealand, promises to complement the European supply by extending the seasonal availability of fresh truffles. In New Zealand we foresee the daily dispatch of harvested truffles via refrigerated air cargo. In the early 1990s, to get some experience of exporting perishable commodities, we sent samples of shoro and porcini, which grow wild in New Zealand, to Germany, Japan, and United Kingdom by refrigerated airfreight. These mushrooms arrived within two days of picking and in very good condition, despite being more perishable than the Périgord black truffle. Subsequent exports of New Zealand Périgord black truffles to the United States and countries in the Pacific Basin arrived in perfect condition.

In other developments, European researchers are also investigating the technical feasibility of applying some advanced preservation techniques to truffles, so that consumers can enjoy long-life fresh truffles that have lost none of their intense flavour. Modified atmosphere packaging and controlled atmosphere storage, irradiation combined with natural bactericide, and edible biofilm packaging technologies are being studied to see if they can provide solutions to the problem of longer-term preservation.

Truffles Without Plants?

Over the past two decades, various claims have been made that it is possible to cultivate the Périgord black truffle away from its host using methods similar to those employed for the cultivation of saprobic mushrooms. One of the first examples came from California, where a product was grown by inoculating trays of material with cultures grown from pieces of Périgord black truffle. The resulting cultures were then harvested and packaged in 14-g bottles labelled La Truffe. The company, California Truffle Co. Ltd., filed for bankruptcy in 1991.

Similar claims were made in 1996, this time in the Japanese press, that it was possible to cultivate the Italian white truffle away from its host by growing the fungus on "sawdust, tofulees, and other material." In 2001 there was also considerable speculation at the Second International Conference on Edible Mycorrhizal Mushrooms over a claim from a Japanese group that it was possible to cultivate all species of edible mycorrhizal mushrooms in the laboratory. We still await a world awash with cheap Japanese-cultured Périgord black and Italian white truffles.

The Future

Decreasing supply and rising prices have provided enormous incentive for research on truffle cultivation. In the past 20 years Périgord black truffle plantations have expanded to cover many thousands of hectares in France, Italy, and Spain, as well as smaller areas outside of Europe. Governments have encouraged truffle cultivation as an alternative source of income for set-aside land and in underdeveloped areas and as a way of reducing agricultural subsidies. However, many questions remain unanswered. With more than 400,000 inoculated trees having been planted each year for more than two decades in France, why has this had no impact on production in that country? Clearly, there are problems. Could it be that research is being aimed in the wrong direction? Some of the onus may rest on those who assess scientists partly on how many papers they publish. Trendy papers in widely read, high-impact journals count for much more than long-term

field studies, which are unlikely to be attractive to these journals, and such valuable long-term studies produce few papers for the time invested.

Science is full of serendipitous findings that make a mockery of targeted or directed research funding. One example was the accidental discovery that compounds produced by fungi can suppress the growth of bacteria. This discovery of penicillin led to a revolution in the treatment of bacterial diseases in humans. Another less-publicized piece of serendipity was the discovery that the bacterium *Staphylococcus pasteuri*, found on the roots of tissue-cultured plants, could inhibit the growth of bianchetto but not *Hebeloma radicosum*, another ectomycorrhizal fungus. Whether *S. pasteuri* commonly occurs in soil is largely irrelevant—what is important is that an organism apparently unrelated to either symbiotic partner might influence the composition of the ectomycorrhizal flora in a soil. This and other interactions between ectomycorrhizal fungi with associated organisms, such as *Rhizobium* bacteria, yeasts, and even other ectomycorrhizal fungi, could be one reason why some ectomycorrhizal fungi are only found in certain areas. Such studies might also explain why only 2 per cent of Italian white truffle mycorrhizas can support fruiting. Fortunately, some new techniques present opportunities for visualizing how organisms interact in the soil. One uses fluorescent markers that can reveal the location of the DNA of specific organisms and thus can be used to see how they interact. Another is the use of microcosms, where ectomycorrhizal fungi can be grown together with their host plant in completely controlled, sterile conditions in which a wide variety of biotic and physical conditions can then be varied.

We are sure that one day someone, somewhere, will develop reliable techniques for cultivating truffles and the many other intractable edible ectomycorrhizal mushrooms. It is our hope, however, that these new techniques will not be methods for cultivating truffles away from their hosts and in factories, as used to cultivate many saprobic mushrooms. Such artificial means would surely toll the bell for an ancient industry that has been touched little by the march of progress and has a special place not just in European hearts but in lovers everywhere of tradition, fine food, and *la grande mystique*.

Appendix 1.
Plants that form arbuscular mycorrhizas and generally can be grown adjacent to truffières without risk of competing fungi

COMMON NAME	BOTANICAL NAME	PLANT FAMILY
Akeake	*Dodonaea viscosa*	Sapindaceae
Akiraho	*Olearia paniculata*	Asteraceae
Almond	*Prunus dulcis*	Rosaceae
Angelica tree	*Aralia*	Araliaceae
Apple	*Malus*	Rosaceae
Apricot	*Prunus armeniaca*	Rosaceae
Ash	*Fraxinus*	Oleaceae
Avocado	*Persea americana*	Lauraceae
Bamboo	*Bambusa*	Pooideae
Banana	*Musa*	Musaceae
Barberry	*Berberis*	Berberidaceae
Bayberry	*Myrica*	Myricaceae
Blackberry	*Rubus fruticosus*	Rosaceae
Black locust	*Robinia*	Fabaceae
Box elder	*Acer negundo*	Aceraceae
Boxwood	*Buxus*	Buxaceae
Broadleaf	*Griselinia*	Griseliniaceae
Buckeye	*Aesculus*	Hippocastanaceae
Burning bush	*Euonymus*	Celastraceae
Cacao	*Theobroma cacao*	Sterculiaceae
Camellia	*Camellia*	Theaceae
Catalpa	*Catalpa*	Bignoniaceae
Cherry	*Prunus avium*	Rosaceae
Chinaberry	*Melia azedarach*	Meliaceae
Coral tree	*Erythrina indica*	Fabaceae
Crab apple	*Malus*	Rosaceae
Cryptomeria	*Cryptomeria japonica*	Taxodiaceae
Cucumber tree	*Magnolia acuminata*	Magnoliaceae
Dogwood	*Cornus*	Cornaceae
Fig	*Ficus carica*	Moraceae
Flax, New Zealand	*Phormium tenax*	Agavaceae
Fuchsia	*Fuchsia*	Onagraceae
Gingko	*Ginkgo biloba*	Ginkgoaceae
Gorse	*Ulex europaeus*	Fabaceae
Grapes	*Vitis*	Vitaceae
Hackberry	*Celtis*	Ulmaceae
Hibiscus	*Hibiscus rosa-sinensis*	Malvaceae
Holly	*Ilex*	Aquifoliaceae

COMMON NAME	BOTANICAL NAME	PLANT FAMILY
Horse chestnut	*Aesculus*	Hippocastanaceae
Juniper	*Juniperus*	Cupressaceae
Kamahi	*Weinmannia racemosa*	Cunoniaceae
Karamu	*Coprosma robusta*	Rubiaceae
Kauri	*Agathis*	Araucariaceae
Korokia	*Corokia buddleoides*	Cornaceae
Kowhai	*Sophora*	Papilionaceae
Lacebark	*Hoheria populnea*	Malvaceae
Lawson cypress	*Chamaecyparis lawsoniana*	Cupressaceae
Lemonwood	*Pittosporum*	Pittosporaceae
Leyland cypress	×*Cupressocyparis leylandii*	Cupressaceae
Macrocarpa	*Cupressus macrocarpa*	Cupressaceae
Magnolia	*Magnolia*	Magnoliaceae
Mahoe	*Melicytus ramiflorus*	Violaceae
Maple	*Acer*	Aceraceae
Marbleleaf	*Carpodetus serratus*	Carpodetaceae
Mulberry	*Morus*	Moraceae
Olive	*Olea europaea*	Oleaceae
Palm	*Cycad*	Cycadaceae
Papaya	*Carica papaya*	Cariceae
Paulownia	*Paulownia*	Paulowniaceae
Peach	*Prunus persica*	Rosaceae
Pear	*Pyrus communis*	Rosaceae
Persimmon	*Diospyros*	Ebenaceae
Plum	*Prunus*	Rosaceae
Podocarp	*Podocarpus*	Podocarpaceae
Pohutukawa	*Metrosideros excelsa*	Myrtaceae
Privet	*Ligustrum*	Oleaceae
Rain tree	*Koelreuteria elegans*	Sapindaceae
Rata	*Metrosideros*	Myrtaceae
Redwood, coastal	*Sequoia sempervirens*	Taxodiaceae
Redwood, giant	*Sequoiadendron giganteum*	Taxodiaceae
Ribbonwood	*Plagianthus betulinus*	Malvaceae
Rowan	*Sorbus*	Rosaceae
Sycamore	*Acer*	Aceraceae
Tree-of-heaven	*Ailanthus altissima*	Simaroubaceae
Tulip tree	*Liriodendron*	Magnoliaceae
Viburnum	*Viburnum*	Caprifoliaceae
Yew	*Taxus*	Taxaceae

Appendix 2.

Plants that can form ectomycorrhizas and may harbour fungi that might compete with truffle fungi. These plants can cause problems if they are present in an area where a truffière is to be established or if they are included in windbreaks.

COMMON NAME	BOTANICAL NAME	PLANT FAMILY
Alder	*Alnus*	Betulaceae
Aspen	*Populus**	Salicaceae
Beech	*Fagus*	Fagaceae
Birch	*Betula*	Betulaceae
Cedar	*Cedrus*	Pinaceae
Cherry, bird	*Prunus padus*†	Rosaceae
Cherry, dwarf	*Prunus cerasus*†	Rosaceae
Cherry, wild	*Prunus avium*†	Rosaceae
Chestnut	*Castanea*	Fagaceae
Douglas fir	*Pseudotsuga menziesii*	Pinaceae
Eucalyptus	*Eucalyptus**	Myrtaceae
Fir	*Abies*	Pinaceae
Hawthorn	*Crataegus*	Rosaceae
Hazelnut	*Corylus*	Betulaceae
Hemlock	*Tsuga*	Pinaceae
Hickory	*Carya*	Juglandaceae
Hornbeam	*Carpinus*	Betulaceae
Ironwood	*Casuarina*	Casuarinaceae
Kanuka	*Kunzea ericoides**	Myrtaceae
Larch	*Larix*	Pinaceae
Lime	*Tilia*	Tiliaceae
Manuka	*Leptospermum scoparium**	Myrtaceae
Oak	*Quercus*	Fagaceae
Pine	*Pinus*	Pinaceae
Poplar	*Populus**	Salicaceae
Redbud	*Cercis canadensis*	Fagaceae
Rockrose	*Helianthemum*	Cistaceae
She-oak	*Casuarina**	Casuarinaceae
Spruce	*Picea*	Pinaceae
Strawberry tree	*Arbutus*	Ericaceae
Walnut	*Juglans*	Juglandaceae
White-leaved rockrose	*Cistus*	Cistaceae
Wild pear	*Pyrus pyraster*	Rosaceae
Wild service tree	*Sorbus torminalis*	Rosaceae
Willow	*Salix**	Salicaceae

Sources: Harley and Harley 1987; De Roman et al. 2005

*Can also form arbuscular mycorrhizas

†Probably is an arbuscular mycorrhizal plant but may occasionally form ectomycorrhizas

Appendix 3.

Some host plants of commercially important species of truffle. The letter "e" indicates the establishment of mycorrhizas under experimental conditions.

FAMILY Host plant species	Mattiro-lomyces terfez-ioides	Tuber aestivum	Tuber borchii	Tuber magna-tum	Tuber melano-sporum	Tuber macro-sporum	Tuber mesen-tericum	Ter-fezia	Tir-mania
BETULACEAE									
Alnus cordata		e	e	e	e				
Betula verrucosa		◆				◆			
CORYLACEAE									
Carpinus betulus		◆			◆				
Corylus avellana		◆	◆	◆	◆	◆	◆		
Corylus colurna		◆					◆		
Corylus heterophylla					◆				
Ostrya carpinifolia		◆	◆	◆	◆	◆	◆		
FABACEAE									
Robinia pseudoacacia	◆							◆	
FAGACEAE									
Castanea sativa		◆	e				◆		
Fagus sylvatica		◆	◆				◆		
Quercus cerris		◆	◆	◆	◆	◆	◆		
Quercus coccifera	◆				◆				
Quercus faginea					◆				
Quercus ilex		◆	◆	◆	◆				
Quercus petraea		◆	◆	◆	◆	◆			
Quercus pubescens		◆	◆	◆	◆	◆	◆		
Quercus robur		◆	◆	◆	◆	◆			
TILIACEAE									
Tilia americana		◆							
Tilia cordata		◆	◆	◆	◆		◆		
Tilia ×europaea		◆	◆	◆	◆	◆	◆		
Tilia platyphyllos		◆	◆	◆	◆	◆	◆		
PINACEAE									
Abies alba		◆		e	e				
Cedrus atlantica, Cedrus deodara		◆	◆	e	e	e			
Larix			◆						
Picea excelsa		◆	◆						
Pinus brutia, Pinus halepensis		◆	◆						

FAMILY Host plant species	Mattirolomyces terfezioides	Tuber aestivum	Tuber borchii	Tuber magnatum	Tuber melanosporum	Tuber macrosporum	Tuber mesentericum	Terfezia	Tirmania
Pinus nigra ssp. *nigra, P. nigra* ssp. *nigricans*		◆	◆				◆		
Pinus pinaster var. *atlantica*		e	◆		e	e		◆	
Pinus pinea		e	◆	e	◆	e	e		
Pinus strobus		e	◆		e	e			
Pinus sylvestris		◆	◆		e	e			
SALICACEAE									
Populus alba			e	◆	e	◆			
Populus nigra		e	◆	◆		◆			
Populus tremula				◆					
Salix alba			e	◆		◆			
Salix caprea			e	◆	e	◆			
CISTACEAE									
Cistus albidus, C. incanus, C. monspeliensis, C. salviaefolius, and other *Cistus*		e	e		◆			◆	
Fumana procumbens	◆								
Helianthemum almeriense, H. guttatum, H. hirtum, H. ledifolium, H. lippii, H. marifolium, H. ovatum, H. salicifolium								◆	◆
Tuberaria								◆	
MIMOSACEAE									
Acacia erioloba, A. hebeclada, and other *Acacia*								◆	

Sources: Ramsbottom 1953; Chevalier et al. 1975; Giovannetti and Fontana 1980–1981; Zambonelli and Branzanti 1984, 1989; Gregori and Tocci 1985; Dexheimer et al. 1985; Bencivenga and Vignozzi 1989; Zambonelli et al. 1989; Pirazzi 1990; Gregori 1991; Zambonelli and Di Munno 1992; Pegler al. 1993; Craddock 1994; Lansac et al. 1995; Astier 1998; Olivier et al. 2002; Iddison 2004; Associazione tartufai bresciani 2005; Centro sperimentale di tartuficoltura 2005; Raggi vivai 2005; British Mycological Society 2006; Truffle UK 2006

Appendix 3

Truffle common names in four languages.

TRUFFLE	ENGLISH	FRENCH	ITALIAN	SPANISH
Mattirolomyces terfezioides	sweet truffle			
Tuber aestivum	Burgundy or summer	truffe de Bourgogne	scorzone or tartufo estivo	trufa de verano
Tuber borchii	bianchetto	truffe blanche de printemps, blanquette	bianchetto or marzuolo	trufa de Borch
Tuber magnatum	Italian white	truffe blanche du Piémont	tartufo bianco pregiato	trufa blanca
Tuber melanosporum	Périgord black	truffe du Périgord	tartufo nero pregiato	trufa negra
Tuber macrosporum	smooth black truffle	truffe lisse	tartufo nero liscio	trufa negra
Tuber mesentericum	Bagnoli truffle	truffe mésentérique or truffe de Bagnoli	tartufo di Bagnoli	tartufo di Bagnoli
Terfezia	desert truffle, Boudier's truffle, Kalahari truffle	truffes du désert	tartufo giallo	turma
Tirmania	desert truffle, Khulas	truffes du désert	tartufo delle sabbie	turma

Appendix 3

Host plant common names (with family names) in four languages.

HOST PLANT	ENGLISH	FRENCH	ITALIAN	SPANISH
BETULACEAE				
Alnus cordata	Italian alder,	aulne de Corse	ontano napoletano	aliso napolitano
Betula verrucosa	silver birch	bouleau verruqueux	betulla	abedul de plata
CORYLACEAE				
Carpinus betulus	European hornbeam	charme communs	carpino bianco	carpe blanco

HOST PLANT	ENGLISH	FRENCH	ITALIAN	SPANISH
Corylus avellana	common hazel-nut, cobnut	noisetier commun	nocciolo	avellano
Corylus colurna	Turkish hazelnut	noisetier de Byzance or noisetier turc	nocciolo di Costantinopoli	avellano turco
Corylus heterophylla	Siberian hazelnut	noisetier du Japon	nocciolo giapponese	avellano de Siberia
Ostrya carpinifolia	hop hornbeam	charme noir	carpino nero	carpe negro
FABACEAE				
Robinia pseudoacacia	black locust	faux-acacia	robinia	falsa acacia
FAGACEAE				
Fagus sylvatica	European beech	hêtre	faggio	haya
Quercus cerris	turkey oak	chêne chevulu	cerro	marojo, roble turco or roble de Turquìa
Quercus coccifera	Palestine oak	chêne kermés	quercia spinosa	coscoja
Quercus faginea	Portuguese oak	chêne zeén	quejigo	
Quercus ilex	holm oak	chêne vert	leccio	encina
Quercus petraea	sessile oak	chêne sissile ou rouvre	rovere	roble albar
Quercus pubescens	pubescent oak	chêne pubescent	roverella	roble pubescente
Quercus robur	common oak	chêne pedonculé	farina	roble común
TILIACEAE				
Tilia americana	American basswood	Tilleul d'Amérique	tiglio Americano	tilo Americano
Tilia cordata	small-leaved lime	tilleul à petite feuilles	tiglio selvatico	tilo de hoja pequeña
Tilia ×europaea	European lime	tilleuls	tiglio comune	tilo común
Tilia platyphyllos	large-leaved lime	tilleul à grande feuilles	tiglio nostrano	tilo de hojas grandes
PINACEAE				
Abies alba	common silver fir	sapin pectine	abete bianco	abeto blanco

HOST PLANT	ENGLISH	FRENCH	ITALIAN	SPANISH
Cedrus atlantica, Cedrus deodara	Atlantic cedar	cèdre de l'Atlas	cedro dell'Atlante	cedro del Atlas
Larix	larch	mélèze	larice	alerce
Picea excelsa	Norway spruce	sapin	Abete rosso	piceas
Pinus brutia, Pinus halepensis	Turkish pine	pin d'Alep	pino d'Aleppo	carrasco
Pinus nigra ssp. *nigra, P. nigra* ssp. *nigricans*	Austrian pine	pin noir d'Autriche	pino nero	pino laricio
Pinus pinaster var. *atlantica*	maritime pine	pin maritime	pino marittimo	pino maritimo
Pinus pinea	stone pine	pin pignon	pino domestico	pino piñonero
Pinus strobus	Weeymouth pine	pin du Lord	pino strobo	pino do Lord
Pinus sylvestris	Scots pine	pin sylvestre	pino silvestre	pino albar
Pseudotsuga menziesii	Douglas fir	sapin de Douglas	douglasia	abeto Douglas
SALICACEAE				
Populus alba	white poplar	peuplier blanc	pioppo bianco	álamo blanco
Populus nigra	black poplar	peuplier noir	pioppo nero	álamo negro
Populus tremula	trambling aspen	tremble	pioppo tremulo	álamo temblón
Salix alba	white willow	saule blanc	salice bianco	sauce blanco
Salix caprea	pussy willow	saule marsault	silicone	sauce cabruno
CISTACEAE				
Cistus albidus, C. incanus, C. monspeliensis, C. salviaefolius, and other *Cistus*	rockrose	ciste	cisto	jara
Fumana procumbens	rockrose	hélianthéme	eliantemo	jarilla

HOST PLANT	ENGLISH	FRENCH	ITALIAN	SPANISH
Helianthemum almeriense, H. guttatum, H. hirtum, H. ledifolium, H. lippii, H. marifolium, H. ovatum, H. salicifolium	rockrose	héliantheme	eliantemo	jarilla
Tuberaria				
MIMOSACEAE				
Acacia erioloba, A. hebeclada, and other *Acacia*	camel thorn	acacia	acacia	acacia

Appendix 4.

Combined production of Périgord black truffle (*Tuber melanosporum*) and winter truffle (*Tuber brumale*) in France and Spain, 1990–2005

SEASON	FRANCE (TONNES)	SPAIN (TONNES)
1990–1991	17	30
1991–1992	20	10
1992–1993	31	23
1993–1994	22	9
1994–1995	12	4
1995–1996	19	20
1996–1997	50	25
1997–1998	30	80
1998–1999	14	7
1999–2000	40	35
2000–2001	35	6
2001–2002	15	20
2002–2003	n.a.	40
2003–2004	n.a.	7
2004–2005	n.a.	22

Sources: Reyna 2005; Michel Courvoisier, personal communication

n.a.: Data not available

Appendix 5.
Truffle production in Italy, 1950–2006

YEAR	JANUARY– MARCH (TONNES)	APRIL– JUNE (TONNES)	JULY– SEPTEMBER (TONNES)	OCTOBER– DECEMBER (TONNES)	TOTAL (TONNES)
1950	n.a.	n.a.	n.a.	n.a.	30.4
1960	n.a.	n.a.	n.a.	n.a.	76.4
1970	n.a.	n.a.	n.a.	n.a.	83.8
1980	25.0	11.9	15.8	19.1	71.8
1990	36.8	8.8	36.6	25.2	107.4
1991	33.5	6.8	14.8	22.1	77.2
1992	20.1	768.5	25.6	32.5	846.7
1993	16.1	8.4	22.5	91.5	138.5
1994	14.0	10.3	17.3	32.4	74.0
1995	17.2	11.4	49.9	117.0	195.5
1996	24.5	1.2	27.3	31.2	84.2
1997	13.5	12.7	11.7	47.8	85.7
1998	16.6	15.2	14.6	18.9	65.3
1999	14.3	15.1	26.8	30.3	86.5
2000	24.4	16.5	36.0	21.0	97.9
2001	16.4	17.3	20.9	14.7	69.3
2002	14.8	18.8	34.1	50.3	118.0
2003†	23.1 / 4.3	19.5 / 2.7	25.8 / 1.4	16.5 / 4.7	84.9 / 13.1
2004†	13.0 / 3.3	14.3 / 0.8	19.7 / 2.2	18.6 / 9.9	65.6 / 16.2
2005†	12.7 / 5.4	17.7 / 1.3	29.0 / 0.5	22.2 / 11.2	81.6 / 18.4
2006†	10.8 / 4.8	13.4 / 1.4	12.7 / 0.4	n.a.	n.a.

Sources: Pettenella et al. 2004; Bollettino mensile di statistica, ISTAT

n.a.: Data not available

†Black truffles (predominantly *Tuber melanosporum* and *T. aestivum*) / white truffles
(*T. magnatum* and *T. borchii*)

Appendix 6.

Some desert truffles

BOTANICAL NAME	COMMON NAMES	DISTRIBUTION	HARVEST
Terfezia arenaria (= T. leonis)	criadilla de tierra, terfez	Algeria, France, Greece, Italy (southern peninsula, Sardinia, Sicily), Kuwait, Libya, Morocco, Portugal, Serbia, Spain, Tunisia, Turkey	April, May
Terfezia boudieri	Boudier's terfez	Algeria, Egypt, Kuwait, Israel, Italy (Sardinia), Libya, Morocco, Spain, Syria, Tunisia, Turkey	March, April
Terfezia claveryi	terfess rouge de Tafilatet, criadilla vaquera, el kamah, fugaa, faqah, fig-aa, kamé, terfas, terfez, torfàs, torfez, turma, zobaïdi, zubadee	Algeria, Egypt, Iran, Iraq, Israel, Kuwait, Libya, Morocco, Saudi Arabia, Spain, Syria, Tunisia	March, April
Terfezia leptoderma		France, Italy (Sardinia), Morocco, Spain	April, May
Kalaharituber pfeilii (= Terfezia pfeilii)	dcoodcoò, godekos, hawan, haban ilhawas, Kalahari truffle, Kalahari aartappel, knol, n'abba, nama, omajowa, omatumbula, tkabba	Botswana, Namibia, South Africa	March to June
Tirmania nivea	fuga, kamaa, kima, chima	Algeria, Egypt, Italy (Sicily but not recently), Kuwait, Libya, Morocco, Spain, Syria, Tunisia	December to the end of March
Tirmania pinoyi	khulasi	Algeria, Egypt, France, Italy, Kuwait, Libya, Morocco, Spain, Tunisia	December to the end of March
Delastria rosea		Italy (Sardinia), Morocco	December, January
Picoa juniperi		Morocco, Spain, Tunisia	spring
Picoa lefebvrei	chivato de la turma	Algeria, Iraq, Kuwait, Libya, Spain, Tunisia	spring
Loculotuber gennadii	terfess	Canary Islands, France, Greece, Italy (Sardinia), Morocco, Spain	end of autumn, end of February, March, April
Tuber oligospermum		France, Israel, Italy, Morocco, Portugal, Spain	autumn in southern Europe, but also spring in hotter climates

Sources: Alsheikh and Trappe 1983; Alvarez et al. 1992; Van der Walt and Le Riche 1999; Montecchi and Sarasini 2000; Moreno et al. 2000, 2005; Khabar et al. 2001, 2005; Mshigeni 2001; Pegler 2002; Iddison 2004a, 2004b; Al-Rasheed 2005; Ferdman et al. 2005

Appendix 7.

Climatic data for various centres adjacent to or with similar climates to Périgord black, Italian white, Burgundy, and bianchetto truffle-producing areas

	PÉRIG-ORD BLACK	ITALIAN WHITE	BUR-GUNDY	BIAN-CHETTO	LATITUDE	ELEVA-TION (M)
SWEDEN						
Gotland			◆		57°40'N	51
DENMARK						
Copenhagen (Schleswig*)			◆	◆	55°36'N	5
UNITED KINGDOM						
Durham			◆	◆	54°40'N	102
Sheffield			◆		53°20'N	131
Stratford upon Avon			◆	◆	52°20'N	49
Lyneham (Wiltshire)			◆	◆	51°30'N	145
Wye (Kent)			◆	◆	51°50'N	56
Yeovilton			◆	◆	51°00'N	20
Eastbourne			◆		50°30'N	7
FRANCE						
Caen			◆	◆	49°18'N	67
Bourges			◆	◆	47°06'N	166
Dijon	◆		◆	◆	47°18'N	227
Clermont-Ferrand	◆		◆	◆	45°50'N	329
Lyon	◆		◆	◆	45°42'N	201
Orange (Marseille*)	◆		◆		44°08'N	60
Toulouse	◆		◆		43°36'N	153
SWITZERLAND						
Lugano, Ticino		◆	◆	◆	46°00'N	273
CROATIA						
Istria (Rijeka, Trieste*)	◆	◆	◆	◆	45°13'N	85
ITALY						
Cuneo (Torino*)	◆	◆	◆	◆	44°55'N	384
Bologna		◆		◆	44°30'N	84
Arezzo	◆	◆	◆	◆	43°28'N	249
Perugia	◆	◆	◆	◆	43°06'N	205
Campobasso	◆	◆	◆	◆	41°36'N	807
Sicily			◆	◆	38°20'N	21

ANNUAL RAINFALL (MM)	ACCUMU-LATED DEGREE DAYS (>10°C)	MEAN DAILY TEMPERA-TURE IN SUMMER JULY/JAN (°C)	MEAN DAILY TEMPERA-TURE IN WINTER JAN/JULY (°C)	ANNUAL SUNSHINE HOURS	SUMMER SUN-SHINE HOURS (APRIL−SEPT/ SEPT−APRIL)
514	570	15.9	−1.1	1579	1148
525	671	16.4	0.1	1597	1197*
643	489	15.2	3.4	1375	932
825	685	16.6	4.0	1381	1001
846	667	16.2	3.8	1379	973
719	676	16.6	3.9	1565	1102
728	756	16.8	4.3	1603	1123
725	751	16.8	4.8	1523	1078
790	895	17.2	5.7	1849	1309
711	852	17.0	4.5	1762	1191
723	1115	19.2	3.3	1827	1279
732	1137	19.7	1.6	1831	1307
563	1102	19.2	2.7	1990	1317
825	1312	20.7	2.6	1975	1411
810	1562	22.0	4.0	2837	1814
655	1513	21.3	5.4	2051	1350
1545	1421	21.3	3.1	2028	1245
1045*	1837	23.0	5.0	2388	1405*
948	1379	22.0	3.5	1989*	1246*
589	2009	24.6	2.4	2064	1458
755	1349	21.0	4.0	n.a.	n.a.
816	1622	22.4	4.4	2061	1457
628	1436	21.5	3.8	2113	1365
611	3125	26.2	12.5	n.a.	n.a.

	PÉRIG-ORD BLACK	ITALIAN WHITE	BUR-GUNDY	BIAN-CHETTO	LATITUDE	ELEVA-TION (M)
SPAIN						
Pamplona, Navarra	◆		◆		43°48'N	449
Perarrua, Huesca (Zaragoza*)	◆		◆		42°16'N	517
Arroyo Cerezo, Valencia	◆		◆		40°07'N	1344
Cuenca, Cuenca (Madrid*)	◆		◆		40°05'N	1001
Santiago de la Espada (Albacete*)	◆				38°20'N	1328
UNITED STATES						
Chapel Hill, N.C. (Washington, D.C.*)	◆				36°10'N	190
Ukiah, CA	◆				39°20'N	189
NEW ZEALAND						
Opotiki	◆				38°00'S	6
Gisborne	◆				38°40'S	9
Taumarunui	◆				38°55'S	171
Paraparaumu	◆				40°55'S	7
Nelson	◆				41°15'S	10
Waipara (Christchurch*)	◆			◆	43°05'S	64
Ashburton	◆			◆	43°54'S	101

Sources: Arléry 1970; Wallén 1970, 1977; Cantù 1977; Delmas 1978; New Zealand Meteorological Service 1980; Reyna 2000; Reyna et al. 2005; Hong Kong Observatory 2007; Met Office 2007; www.eurometeo.com/english/climate, www.met-office.gov.uk/climate/uk, www.hko.gov.hk/wxinfo/climat/world/eng/world_climat_e.htm, www.wunderground.com, www.histrica.com/istra-topics/1-locale-en.html

*Data within the row are from this nearby location.

n.a.: Data not available

ANNUAL RAINFALL (MM)	ACCUMU-LATED DEGREE DAYS (>10°C)	MEAN DAILY TEMPERA-TURE IN SUMMER JULY/JAN (°C)	MEAN DAILY TEMPERA-TURE IN WINTER JAN/JULY (°C)	ANNUAL SUNSHINE HOURS	SUMMER SUN-SHINE HOURS (APRIL–SEPT/ SEPT–APRIL)
963	1342	20.0	4.5	n.a.	n.a.
644	1479	21.7	2.9	2638*	1689*
552	749	18.9	3.1	n.a.	n.a.
569	1352	21.7	3.1	2678*	1714*
673	1547	23.2	3.3	2705*	1713*
1054	2341	25.5	3.5	2173*	1545*
943	2341	22.7	7.5	n.a.	n.a.
1400	1493	18.5	9.2	2169	1227
1058	1430	18.3	9.0	2172	1283
1443	1292	18.3	7.9	1704	1079
1054	1167	17.1	8.3	2043	1227
986	1038	17.2	6.5	2397	1377
729	1049	17.5	6.5	1999*	1175*
757	896	16.5	5.2	1892	1092

Appendix 8.

Physical and chemical properties of Italian white truffle (*Tuber magnatum*) soils

	RANGE OR MEAN ± SD
pH (in water)	7.3–8.1
Bulk density (g/ml)	1.06 ± 0.12
Total calcium (per cent)	2.3–13.9
Extractable calcium (μg/ml)	2000–5000
Extractable phosphorus (μg/ml)	4.0–42
Extractable potassium (μg/ml)	80–620
Extractable magnesium (μg/ml)	135–625
Extractable sodium (μg/ml)	2.0–26
Extractable sulphur (μg/ml)	2.0–25
Extractable boron (μg/ml)	0.1–1.3
Extractable iron (μg/ml)	40–117
Organic carbon (per cent)	0.8–8.8
Carbon-to-nitrogen ratio	10–23.3
Cation exchange capacity (mmol/kg)	1.72 ± 0.47

Sources: Lulli et al. 1991, 1992, 1993; Hall et al. 1998

Appendix 9.
Climatic features in Gard, Drôme, and Vaucluse and
production figures for Carpentras, France, 1987–1993

YEAR	PRODUCTION (TONNES)	ASSESS- MENT OF PRODUC- TION	CLIMATIC FEATURES	NOTES
1987	80	good	hot and humid, stormy	Abundant rain alternating with periods of relative dryness not exceeding 20 days. Rise and fall of soil temperature of 5°C over a seven-day period in May.
1988	30	medium	fairly hot and humid	Average rainfall but somewhat less than normal in July. Soil temperature not oscillating as much as in 1987.
1989	15	very bad	cold and dry	Prolonged cold weather from the end of April until mid-June, with soil temperatures not reaching 12°C until 10 June combined with 100 days of drought from May to August.
1990	20	bad	cold and dry in spring	
1991	15	very bad	prolonged winter drought	
1992	20	bad	cold and dry in spring	Sufficient rainfall from May to September, but 50 days of prolonged cold soil temperatures from 1 June to 14 July—the soil temperature was only 12.5°C on 8 July, well below the average of 19°C.
1993	30	medium	fairly hot and wet	

Source: Bardet and Fresquet 1995

Appendix 10.

Factors that might trigger fruiting or affect the productivity of truffières. Some of these factors will interact with or affect others as well.

FACTOR	POTENTIAL TRIGGER	POSSIBLE EFFECT ON YIELD	INTERACT WITH FACTORS
1 Soil temperature	✦		2, 3, 4, 5, 7, 8, 9, 10, 16, 17
2 Soil moisture and irrigation	✦	✦	1, 3, 4, 7, 10, 13,15
3 Pruning	✦		1, 2, 4, 6, 8, 9 10, 13, 15
4 Plant density, spacing, orientation, and amount of sunlight striking the ground	✦	✦	1, 2, 3, 5, 8, 9, 10, 13, 14, 15
5 Presence of beneficial organisms	✦	✦	8, 9, 15
6 Carbohydrate flow to roots	✦	✦	5, 8, 9, 14, 15
7 Mulches and weeds	✦	✦	1, 2, 5, 8, 9, 10, 11, 12, 15, 16, 17
8 Amount of truffle mycelium in the soil	✦	✦	5, 9
9 Competing fungi	✦	✦	5, 8
10 Soil aeration	✦	✦	1, 2, 5, 8, 9, 15
11 Major and minor nutrients	✦	✦	5, 8, 9, 13, 15
12 Trace elements	✦	✦	5, 8, 9, 13, 15
13 Host plant		✦	4, 5, 6, 8, 9, 14, 15
14 Pests		✦	13, 16, 17
15 Diseases of the host plant and symbiont		✦	13, 16, 17
16 Elevation		✦	
17 Latitude		✦	

Appendix 11.
Manganese fertilizers

COMPOUND	FORMULA	PERCENT MANGANESE	COMMENTS
Manganese sulphate	$MnSO_4 \cdot H_2O$	24–30	soluble, quick-acting
Manganese chloride	$MnCl_2$	17	soluble, quick-acting
Manganese carbonate	$MnCO_3$	31	insoluble, slow-acting
Manganese chelate	$Na_2MnEDTA$	5–12	quick-acting
Manganese oxide	MnO_2	40	insoluble, slow-acting

Appendix 12.
Commonly used boron fertilizers

FERTILIZER	FORMULA	PERCENT BORON
REFINED PRODUCTS		
Sodium tetraborate decahydrate	$Na_2B_4O_7 \cdot 10H_2O$	11.3
Sodium tetraborate pentahydrate	$Na_2B_4O_7 \cdot 5H_2O$	14.9
Boric acid	$B(OH)_3$	17.5
Solubor	$Na_2B_8O_{13} \cdot 4H_2O$	20.8
Sodium tetraborate	$Na_2B_4O_7$	21.4
CRUSHED ORES		
Ascharite	$2MgO \cdot B_2O_3 \cdot H_2O$	variable
Colemanite	$2CaO \cdot 3B_2O_3 \cdot 5H_2O$	variable
Datolite	$2CaO \cdot B_2O_3 \cdot 2SiO_2 \cdot H_2O$	variable
Hydroboracite	$CaO \cdot MgO \cdot 3B_2O_3 \cdot 6H_2O$	variable
Ulexite	$Na_2O \cdot 2CaO \cdot 5B_2O_3 \cdot 5H_2O$	variable

Source: Shorrocks 1997

Appendix 13.

The authorized harvesting dates for truffles in France and Italy. In France fresh truffles can be sold for another 10 days after the close of a season.

BOTANICAL NAME	COMMON NAME	FRANCE	ITALY
Tuber magnatum	Italian white truffle		1 Oct–31 Dec
Tuber melanosporum	Périgord black truffle	1 Dec–31 March	15 Nov–15 March
Tuber aestivum	summer truffle	1 May–31 Sept	1 May–30 Nov
	Burgundy truffle		1 Oct–31 Dec
Tuber borchii	bianchetto		15 Jan–30 April
Tuber brumale	winter black truffle	1 Dec–31 March	1 Jan–15 March
Tuber brumale var. moschatum	musky truffle		15 Nov–15 March
Tuber macrosporum	smooth black truffle		1 Sept–31 Dec
Tuber mesentericum	Bagnoli truffle	1 Sept–31 Dec	1 Sept–31 Jan
Tuber indicum	Chinese truffle	1 Dec–31 March	

Sources: See Chevalier and Frochot (1997) for a summary of French regulations. Legge quadro nazionale 752/1985 and 162/1991; Sainte Alvere 2005

TRUFFLE ORGANIZATIONS AND PLANT AND EQUIPMENT SUPPLIERS

Australian Truffle Growers
  Association
PO Box 7426
Sutton NSW 2620
Australia
www.trufflegrowers.com.au

Fédération Française des
  Trufficulteurs
7 Bis Rue du Louvre
75001 Paris
France
Tel. +33 142 360 329
Fax +33 142 362 693
www.fft-tuber.org

Kungming Yunri Foods Co. Ltd.
Room 311
Yunnan Business Hotel
139 Dong Feng Xi Road
Kunming
Yunnan 650041
China
Tel. +86 871 362 6046
Fax +86 871 362 6092
www.sinohost.com/yunnan_pages/
  mushrooms

La Trufficulteur Français
7 Rue du Jardin Public
BP 7065
24007 Périgueux
France
Tel. +33 142 360 329
Fax +33 142 362 693
www.fft-tuber.org/bulletin.asp

New Zealand Truffle Association
PO Box 10629
Wellington 6143
New Zealand
Tel. +64 4 473 6040
Fax +64 4 473 6041
www.southern-truffles.co.nz

Pébeyre S.A.
66 Rue Frédéric-Suisse
46000 Cahors
France
Tel. +33 565 222 480
Fax +33 565 300 166
www.pebeyre.fr

Rougié
Avenue du Périgord
BP 118
24203 Sarlat
France
Tel. +33 553 317 200
Fax +33 553 594 086
www.foie-gras-rougie.com

Tofani Tartufi di Tofani Stefania
Via Bellaria 37
61041 Acqualagna (PU)
Italy
Tel. and Fax +39 0721 798 918
E-mail tartufi.tofani@virgilio.it

Urbani Tartufi
S. Anatolia di Narco
06040 Perugia (PG) Umbria
Italy
Tel. +39 0743 613 171
Fax +39 0743 613 035
www.urbani.com

Soil Testing Laboratories
Laboratoire d'analyses
    Agro-Environmentales
LCA La Rochelle
1 Rue Samuel Champlain
17974 La Rochelle
Cedex 9
France
Tel. +33 546 434 545
Fax +33 546 675 680
www.laboratoirelca.com

New World Truffières
Charles Lefevre
PO Box 5802
Eugene, Oregon 97405
United States
Tel. +1 541 513 4176
www.truffletree.com

SWEP Analytical Laboratories
45–47/174 Bridge Road
Keysborough
PO Box 583
Noble Park VIC 3174
Australia
Tel. +61 39701 6007
Fax +61 39701 5712
www.swep.com.au

Truffles & Mushrooms (Consulting)
    Ltd
P.O. Box 268
Dunedin 9054
New Zealand
Tel. +64 3 454 3574
www.trufflesandmushrooms.co.nz

Truffles UK
P.O. Box 5389
Cattistock
Dorchester
Dorset DT2 0XN
United Kingdom
Tel. +44 193 583 819
Fax +44 193 583 820
www.truffle-uk.co.uk

Infected Trees
Agri-Truffe
Domaine de Lalanne
33490 Saint Maixant
France
Tel. +33 556 620 053
Fax +33 556 620 963
www.agritruffe.eu

New World Truffières
Charles Lefevre
PO Box 5802
Eugene, Oregon 97405
United States
Tel. +1 541 513 4176
www.truffletree.com

Oakland Truffiére
Ferry Road
RD
Gisborne
New Zealand
Tel. +64 6862 5597
Fax +64 6862 5598
www.oakland-truffles.com

Raggi Vivai
Via Cerchia di S. Egidio 3000
47023 Cesena (FC)
Italy
Tel. +39 0547 382 171
Fax +39 0547 631 874
www.raggivivai.it

Robin Pépinières EARL
Le Village
05500 Saint Laurent du Cros
France
Tel. +33 492 504 316
Fax +33 492 504 757
www.robinpepinieres.com

Symbiotic Systems NZ Ltd.
P.O. Box 7116
Dunedin 9040
New Zealand
Tel. +64 3 453 4948
Fax +64 3 453 4945
www.trufflesandmushrooms.co.nz

Truffles & Mushrooms (Consulting)
    Ltd
P.O. Box 268
Dunedin 9054
New Zealand
Tel. +64 3 454 3574
www.trufflesandmushrooms.co.nz

Machinery and Equipment
Agrigarden Distributors
24-b Morrin Road
Mt Wellington 1006
Auckland, New Zealand
Tel. +64 9574 5365
Fax +64 9574 5364
www.agrigarden.co.nz

KBC Tree Shelters
Killyleagh Box Company Ltd.
39 Shrigley Road
Killyleagh, Co. Down BT30 9SR
Northern Ireland
Tel. +44 28 4482 8708
Fax +44 28 4482 1222
www.killyleaghbox.co.uk/
    horticulture.htm

PowerFarming
86 Thames Street
PO Box 6
Morrinsville, New Zealand
Tel. +64 7902 2200
Fax +64 7902 2201
www.powerfarming.co.nz

Tubex Ltd.
Aberaman Park
Aberdare CF44 6DA
United Kingdom
Tel. +44 1685 888 000
www.tubex.com

REFERENCES

P LEASE SEE THE Truffles & Mushrooms Web site (www.trufflesandmush rooms.co.nz) for a list of the following references and where they relate to the text. The location of information on the World Wide Web is constantly changing. If a Web reference is out of date, go to the home page and try the search option if there is one or the site map. If this fails, try a good search engine such as Google.

Acqualagna, capitale del tartufo. 2002. Borsa del tartufo. Available via www. acqualagna.info/modules.php?name=Content&pa=showpage&pid=82.

Agate, E. 2003. *Woodlands*. BTCV Ltd., Doncaster, U.K. Available via www.handbooks.btcv.org.uk/handbooks/content/chapter/682 and /690.

Agerer, R. 1990. Studies of ectomycorrhizae. XXIV. Ectomycorrhizae of *Chroogomphus helveticus* and *C. rutilis* (Gomphidaceae, Basidiomycetes) and their relationships to those of *Suillus* and *Rhizopogon*. *Nova Hedwigia* 50:1–63.

Agerer, R. 1995. Anatomical characteristics of identified mycorrhizas: an attempt towards a natural classification. In: Varma, A.K., Hock, B., eds. *Mycorrhiza Structure, Function, Molecular Biology and Biotechnology*. Springer Verlag, Berlin. Pp. 685–734.

Agerer, R., Rambold, G. 1996. DEEMY: a DELTA-based system for characterization and determination of ectomycorrhizae (ver. 1.0). CD-ROM. Section Mycology, Institute for Systematic Botany, University of Munich.

Agribusinessonline. 2003. Truffles, fresh or chilled, U.S. imports for consumption, annual data. Available via www.agribusinessonline.com/stats/07095200_truffles. asp.

Agri Truffe. 2006. Parce que votre réussite est la nôtre! Available via www.agritruffe. com.

Ahmad, S. 1956. *Fungi of West Pakistan*. Biological Society of Pakistan Monograph no. 1. Biological Society of Pakistan, Lahore.

Alloway, B. 2002. Zinc: the vital micronutrient for healthy, high-value crops. Available via www.initiative-zink.de/163.htm.

Al-Rasheed, M.T. 2005. Of desert truffles and regulations. Arab News. Available via www.arabnews.com/?page=7&section=0&article=58115&d=27&m=1&y=2005&pix=opinion.jpg&category=Opinion.

Al-Rasheed, M.T. 2005. Of desert truffles and regulations. Available via http://xrdarabia.org/blog/archives/2005/01/26/of-desert-truffles-and-regulations.

Al-Ruqaie, I.M. 2002. Effect of different treatment processes and preservation methods on the quality of truffles. 1. Conventional methods (drying/freezing). *Pakistan Journal of Biological Sciences* 5:1088–1093.

Alsheikh, A.M. 1994. Taxonomy and mycorrhizal ecology of the desert truffles in the genus *Terfezia*. Ph.D. diss., Oregon State University, Corvallis.

Alsheikh, A.M., Trappe, J.M. 1983. Desert truffles: the genus *Tirmania*. *Transactions of the British Mycological Society* 81:83–90.

Alvarez, I.F., Parladé, J., Trappe, J.M. 1992. *Loculotuber gennadii* gen. et comb. nov. and *Tuber multimaculatum* sp. nov. *Mycologia* 84:925–929.

Ambra, R., Grimaldi, B., Zamboni, S., Filetic"i, P., Macino, G., Ballario, P. 2004. Photomorphogenesis in the hypogeous fungus *Tuber borchii*: isolation and characterization of Tbwc-1, the homologue of the blue-light photoreceptor of *Neurospora crassa*. *Fungal Genetics and Biology* 41:6888–6897.

American Phytopathological Society. 2006. www.shopapspress.org/index.html.

Amicucci, A., Guidi, C., Zambonelli, A., Potenza, L., Stocchi, V. 2000. Multiplex PCR for the identification of white *Tuber* species. *FEMS Microbiology Letters* 189:265–269.

Amicucci, A., Guidi, C., Zambonelli, A., Potenza, L., Stocchi, V. 2002a. Molecular approaches for the detection of truffle species in processed food products. *Journal of Science and Food Agriculture* 82:1391–1397.

Amicucci, A., Potenza, L., Guidi, C., Rossi, C., Bertini, L., Zambonelli, A., Stocchi, V. 2002b. Molecular techniques in the study of edible ectomycorrhizal mushrooms. In: Hall, I.R., Wang, Y., Danell, E., Zambonelli, A., eds. *Edible Mycorrhizal Mushrooms and Their Cultivation: Proceedings of the Second International Conference on Edible Mycorrhizal Mushrooms*. Christchurch, New Zealand, 3–5 July 2001. CD-ROM. New Zealand Institute for Crop and Food Research Limited, Christchurch.

Amicucci, A., Rossi, I., Potenza, L., Zambonelli, A., Agostini, D., Palma, F., Stocchi, V. 1996. Identification of ectomycorrhizae from *Tuber* species by RFLP analysis of the ITS region. *Biotechnology Letters* 18(7):821–826.

Amicucci, A., Zambonelli, A., Giomaro, G., Potenza, L., Stocchi, V. 1998. Identification of ectomycorrhizal fungi of the genus *Tuber* by species-specific ITS primers. *Molecular Ecology* 7:273–277.

Amicucci, A., Zambonelli, A., Guidi, C., Stocchi, V. 2001. Morphological and molecular characterization of *Pulvinula constellatio* ectomycorrhizae. *FEMS Microbiology Letters* 194:121–125.

Angeletti, M., Landucci, A., Contini, M., Bertuccioli, M. 1990. Caratterizzazione

dell'aroma del tartufo mediante l'analisi gas cromatografica dello spazio di testa. In: Bencivenga, M., Granetti, B., eds. *Atti del secondo congresso internazionale sul tartufo*. Spoleto, Italy, 24–27 November 1988. Comunità Montana dei Monti Martani e del Serano, Spoleto. Pp. 505–509.

Anonymous. 1989. A start-up's magic mushroom. *The Scientist* 3(12):8.

Anonymous. 1993. Black truffles make local debut. *Food Technology in New Zealand* September: 28.

Anonymous. 1995. Electronic sniffer stalks famous fungus. *New Scientist* 7 January: 17.

Anonymous. 1996. Local scientists grow truffles in planter boxes for first time. Kansai Window, *Kippo News* 3:110.

Anonymous. 2001. NZ ideal for truffles. *Horticultural News* September: 7.

Anonymous. 2002. Truffe ou pas truffe? Les techniques d'identification se développent. Direction génerale de la concurrence de la consummation et de la répression des frauds. Available via www.finances.gouv.fr/DGCCRF/02_actualite/breves/brv0403i.htm?ru=02.

Anonymous. 2003. Gourmet gangs get a taste for the finer things. *The Age* (Melbourne), 31 January. Available via www.theage.com.au/articles/2003/01/30/1043804461408.html.

Anonymous. 2004. Truffle growers hope for even bigger crop. Australian Broadcasting Corporation, 28 September. Available via www.abc.net.au/southwestwa/news/200409/s1208712.htm.

Arléry, R. 1970. The climate of France, Belgium, The Netherlands, and Luxembourg. In: Wallén, C.C., ed. *World Survey of Climatology*. Vol. 5, *Climates of Northern and Western Europe*. Elsevier, Amsterdam. Pp. 135–193.

Arnolds, E. 1992. The analysis and classification of fungal communities with special reference to macrofungi. In: Winterhoff, W., ed. *Fungi in Vegetation Science*. Kluwer, Dordrecht. Pp. 7–47.

Arotz. 2006. Trufa: la finca: video clips. Available via www.arotz.com/html/index_prod.htm.

Associazione micologica Bresadola. 2006. Leggi regionali, provinciali, regolamenti, etc. Available via www.ambbresadola.it/Leggi/ItaliaLex.htm (unofficial web site).

Associazione tartufai bresciani. 2005. www.asstartufaibresciani.it/aest_fun.htm.

Associazione telematica tartufai Italiani. 2005. I tartufi in altri paesi e continenti. Available via www.trovatartufi.com/Demo/Il_tartufo/tartufi_in_altro_paesi.htm.

Astier, J. 1998. *Truffes Blanches et Noires: Tuberaceae and Terfeziaceae*. Joseph Astier, La Penne sur Huveaune.

Baciarelli-Falini, L., Rubini, A., Riccioni, C., Paolocci, F. 2006. Morphological and molecular analyses of ectomycorrhizal diversity in a man-made *T. melanosporum* plantation: description of novel truffle-like morphotypes. *Mycorrhiza* 16:475–484.

Balaji, B., Poulin, M.-J., Vierling, H., Piche, Y. 1995. Response of an arbuscular mycorrhizal fungus, *Gigaspora margarita*, to exudates and volatiles from the Ri T-DNA transformed roots of nonmycorrhizal and mycorrhizal mutants of *Pisum sativum* L. Sparkle. *Experimental Mycology* 19:275–283.

Barbieri, E., Gioacchini, A.M., Zambonelli, A., Bertini, L., Stocchi, V. 2005. Determination of microbial volatile organic compounds from *Staphylococcus pasteuri*

against *Tuber borchii* using solid-phase microextraction and gas chromatography/ion trap mass spectrometry. *Rapid Communications in Mass Spectrometry* 19:3411–3415.

Barberi, E., Guidi, C., Bertaux, J., Frey-Klett, P., Garbaye, J., Ceccaroli, P., Saltarelli, R., Zambonelli, A., Stocchi, V. 2007. Occurrence and diversity of bacterial communities in *Tuber magnatum* Pico during truffle maturation. *Environmental Microbiology* 9:2234–2246.

Bardet, M.-C., Fresquet, C. 1995. Influence de la pluviométrie et de la température du sol. *Infos-CTIFL* 110:38–41.

Baring-Gould, S. 1894. Truffles and truffle hunters. In: *The Deserts of Southern France*. Methuen, London.

Barnard, P., ed. 1998. *Biological Diversity in Namibia: A Country Study*. Namibian National Biodiversity Task Force, Directorate of Environmental Affairs, Windhoek.

Bauhin, G. 1623. *Pinax theatri botanici*. Basel.

Baum, C., Schmid, K., Makeschin, F. 2000. Interactive effects of substrates and ectomycorrhizal colonization on growth of a poplar clone. *Journal of Plant Nutrition and Soil Science* 163:221–226.

Bay Gourmet. 1998. Truffle FAQ: section II, basic truffle facts. Available via www.members.tripod.com/~BayGourmet/trufflebas.html.

Beech, H. 2005. Truffle kerfuffle. *Time*, 21 February, 45–47. Available via www.time.com/time/asia/magazine/article/0,13673,501050221-1027587,00.html.

Bellesia, F., Pinetti, A., Bianchi, A., Tirillini, B. 1996. Volatile compounds of the white truffle (*Tuber magnatum* Pico) from middle Italy. *Flavour and Fragrance Journal* 11:239–243.

Belloli, S., Bologna, F., Gregori, G., Zambonelli, A. 2001. Il tartufo nero di Fragno (*Tuber uncinatum* Chatin): ecologia e coltivazione. In: *Proceedings of the Fifth International Congress on the Science and Cultivation of Truffles*. Aix-en-Provence, France, 3–6 March 1999. Federation Française des Trufficulteurs, Paris. Pp. 367–371.

Bencivenga, G., Ferrara, A., Fontana, A., Govi, G., Granetti, G., Gregori, G., Lo Bue, G., Palenzona, M., Rebaudengo, E., Tocci, A., Zambonelli, A. 1987. Valutazione dello stato di micorrizazione di piantine tartufigene. Proposta di un metodo. Ministero Agricoltura e Foreste Corpo Forestale dello Stato, Roma.

Bencivenga, M. 1982. Alcune metodiche di micorrizazione di piante forestali con il tartufo nero pregiato di Norcia o di Spoleto (*Tuber melanosporum* Vitt.). *L'informatore agrario* 38(21):21154–21163.

Bencivenga, M. 1998. Ecology and cultivation of *Tuber magnatum* Pico. In: Danell, E., ed. *Proceedings of the First International Meeting on Ecology, Physiology, and Cultivation of Edible Mycorrhizal Mushrooms*. Uppsala, Sweden, 3–4 July 1998. Available via www.icom2.slu.se/ABSTRACTS/Bencivenga.html.

Bencivenga, M. 2004. Stato attuale della tartuficoltura italiana. In: Bencivenga, M., Donnini, D., Gobbini, A., eds. *Seminario sullo stato attuale della tartuficoltura italiana*. Spoleto e Norcia, 21–22 February 2004. Pp. 9–11.

Bencivenga, M., Granetti, B., eds. 1990. *Atti del secondo congresso internazionale sul*

tartufo. Spoleto, Italy, 24–27 November 1988. Comunità Montana dei Monti Martani e del Serano, Spoleto.

Bencivenga, M., Urbani, G. 1996. Produzione di tartufo bianchetto in una tartufaia coltivata di tre anni. *L'informatore agrario* 52:25–26.

Bencivenga, M., Vignozzi, G. 1989. I tartufi in Toscana. Vantaggio, Firenze.

Bending, G.D., Read, D.J. 1997. Lignin and soluble phenolic degradation by ectomycorrhizal fungi. *Mycological Research* 101:1348–1354.

Bennett, W.F. 1993. *Nutrient Deficiencies and Toxicities in Crop Plants.* American Phytopathological Society, St. Paul, Minn.

Berch, S., ed. 2004. *Proceedings of the Third International Meeting on the Ecology, Physiology and Culture of Edible Mycorrhizal Mushrooms.* Victoria, 15–22 August 2003. 2 CD-ROMs. University of Victoria, Victoria, B.C.

Bertault, G., Raymond, M., Berthomieu, A., Callot, G., Fernandez, D. 1998. Trifling variation in truffles. *Nature* 394:734.

Bertini, L., Agostini, D., Potenza, L., Rossi, I., Zeppa, S., Zambonelli, A., Stocchi, V. 1998a. Molecular markers for the identification of the ectomycorrhizal fungus *Tuber borchii. New Phytologist* 139:565–570.

Bertini, L., Potenza, L., Zambonelli, A., Amicucci, A., Stocchi, V. 1998b. Restriction fragment length polymorphism species-specific patterns in the identification of white truffles. *FEMS Microbiology Letters* 164:397–401.

Bertini, L., Rossi, I., Zambonelli, A., Amicucci, A., Sacchi, A., Cecchini, M., Gregori, G., Stocchi, V. 2006. Molecular identification of *Tuber magnatum* ectomycorrhizae in the field. *Microbiological Research* 161:59–64.

Beuchat, L.R., Brenneman, T.B., Dove, C.R. 1993. Composition of the pecan truffle (*T. texense*). *Food Chemistry* 46:189–192.

Biofilm Institute. 2006. www.biofilm.org.

Blake, A., Crewe, Q. 1978. *Great Chefs of France.* Artists House/Mitchell Beazley, London.

Boa, E. 2004. *Wild Edible Fungi: A Global Overview of Their Use and Importance to People.* Non-Wood Forest Products Report no. 17. Food and Agriculture Organization of the United Nations, Rome. Available via www.fao.org/docrep/007/y5489e/y5489e00.htm.

Bonet, J.A., Oliach, D., Colinas, C. 2004. Cultivo de trufa negra (*Tuber melanosporum*). Available via http://labpatfor.udl.es/docs/cultivotrufa.html.

Bonet, J.A., Fischer, C., Colinas, C. 2006. Cultivation of black truffle to promote reforestation and land-use stability. *Agronomy for Sustainable Development* 26:69–76.

Boudier, E. 1876. Du parasitisme probable de quelques espèces du genre *Elaphomyces* et de la recherché de ces Tubéracées. *Bulletin société botanique de France* 23:115–119.

Boulianne, R.P., Liu, Y., Aebi, M., Lu, B.C., Ku, U. 2000. Fruiting body development in *Coprinus cinereus*: regulated expression of two galectins secreted by a non-classical pathway. *Microbiology* 146:1841–1853.

Brasier, C. 2003. Phytophthoras in European forests: their rising significance. In: Sudden Oak Death Online Symposium. Available via www.apsnet.org/online/proceedings/sod/papers/brasier/default.htm.

Brasier, C.M., Beales, P.A., Kirk, S.A., Denman, S., Rose, J. 2005. *Phytophthora kernoviae* sp. nov., an invasive pathogen causing bleeding stem lesions on forest trees and foliar necrosis of ornamentals in the UK. *Mycological Research* 109:853–859.

Bratek, Z., Gogan, A., Halasz, K., Bagi, I., Erdei, V., Bujaku, G. 2006. The northeast habitats of *Tuber magnatum* known from Hungary. In: Khabar, L., ed. *Le premier symposium sur les champignons hypogés du basin Méditerranéen*. Rabat, Morocco, 5–8 April 2004. Pp. 256–264.

Bratek, Z., Papp, L., Merkl, O. 2001. Beetles and flies living on truffles. In: *Proceedings of the Fifth International Congress on the Science and Cultivation of Truffles*. Aix-en-Provence, France, 3–6 March 1999. Federation Française des Trufficulteurs, Paris. Pp. 191–192.

Breitenbach, J., Kränzlin, F. 1984–1995. *Fungi of Switzerland*. Vols. 1–4. Verlag Mykologia, Lucerne.

Brenneman, T. 2003. Pecan truffle fact sheet. Available via www.interests.caes.uga.edu/pecantruffles/factsheet.htm.

Brillat-Savarin, J.A. 1825. The physiology of taste, or, meditations on transcendental gastronomy. English translation available via www.etext.library.adelaide.edu.au/b/brillat/savarin/b85p/index.html.

British Broadcasting Corporation. 2004a. Puppy's truffle find stuns chef. Available via www.news.bbc.co.uk/1/hi/england/beds/bucks/herts/3441537.stm.

British Broadcasting Corporation. 2004b. Truffles batch 'a huge discovery'. Available via www.news.bbc.co.uk/1/hi/england/berkshire/3668228.stm.

British Broadcasting Corporation. 2005. Italian truffle fetches top price. Available via www.news.bbc.co.uk/2/hi/europe/4434168.stm.

British Mycological Society. 2006. Fungi of the British Isles checklist: genus/species data. Available via http://194.203.77.76/fieldmycology/GBCHKLST/gbsyns.asp?intGBNum=3731 and http://194.203.77.76/fieldmycology/bmsfrd/bmsrecord.asp?intGBNum=9342.

Brown, D. 1998. The effect of applied lime and phosphorus on the competitiveness of *Tuber melanosporum* and other ectomycorrhizal fungi found in Tasmania. Ph.D. diss., University of Tasmania, Hobart.

Brown, D. 2001. The trufficulture in Tasmania. In: *Proceedings of the Fifth International Congress on the Science and Cultivation of Truffles*. Aix-en-Provence, France, 3–6 March 1999. Federation Française des Trufficulteurs, Paris. Pp. 331–333.

Brundrett, M.C. 1991. Mycorrhizas in natural ecosystems. In: Macfayden, A., Begon, M., Fitter, A.H., eds. *Advances in Ecological Research*. Vol. 21. Academic Press, London. Pp. 171–313.

Brundrett, M.C. 2002. Coevolution of roots and mycorrhizas of land plants. Tansley review no. 134. *New Phytologist* 154:275–304.

Brundrett, M.C., Bougher, N., Dell, B., Grove, T., Malajczuk, N. 1996. *Working with Mycorrhizas in Forestry and Agriculture*. ACIAR Monograph no. 32. Australian Centre for International Agricultural Research, Canberra.

Bruni, F. 1891. *Tartufi*. Hoepli, Milano.

Bruns, T.D., Bidartondo, M.I., Taylor, D.L. 2002. Host specificity in ectomycor-

rhizal communities: What do the exceptions tell us? *Integrative and Comparative Biology* 42:352–359.

Bulliard, P. 1791. *Histoire des champignon de la France*. Paris.

Buno, M., Botondi, R., Carlini, P., Massantini, R., Mencatelli, F. 2001. Use of modified atmosphere for storing fresh truffles (*Tuber aestivum* Vitt.). In: *Proceedings of the Fifth International Congress on the Science and Cultivation of Truffles*. Aix-en-Provence, France, 3–6 March 1999. Federation Française des Trufficulteurs, Paris. Pp. 138–139.

Busse, M.D., Fiddler, G.O., Ratcliff, A.W. 2004. Ectomycorrhizal formation in herbicide-treated soils of differing clay and organic matter content. *Water, Air and Soil Pollution* 152:23–34.

Bustan, A., Ventura, Y., Kagan-Zur, V., Roth-Bejerano, N. 2003. Optimizing growing conditions towards intensive cultivation of Périgord black truffles. In: *Proceedings of the Third International Meeting on the Ecology, Physiology and Culture of Edible Mycorrhizal Mushrooms*. Victoria, 15–22 August 2003. 2 CD-ROMs. University of Victoria, Victoria, B.C.

Buzzini, P., Gasparetti, C., Turchetti, B., Cramarossa, M.R., Vaughan-Martini, A., Martini, A., Pagnoni, U.M., Forti, L. 2005. Production of volatile organic compounds (VOCs) by yeasts isolated from the ascocarps of black (*Tuber melanosporum* Vitt.) and white (*Tuber magnatum* Pico) truffles. *Archives of Microbiology* 184:187–193.

Cagniart, P. 1968. Commerce et réglementation de la truffe. In: *Atti del primo congresso internazionale sul tartufo*. Spoleto, Italy, 24–27 May 1968. Ente Rocca di Spoleta, Parma. Pp. 67–72.

Cairney, J.W.G., Meharg, A.A. 2002. Interactions between ectomycorrhizal fungi and soil saprotrophs: implications for decomposition of organic matter in soils and degradation of organic pollutants in the rhizosphere. *Canadian Journal of Botany* 80:803–809.

Cantù, V. 1977. The climates of France, Belgium, The Netherlands, and Luxembourg. In: Wallén, C.C., ed. *World Survey of Climatology*. Vol. 6, *Climates of Central and Southern Europe*. Elsevier, Amsterdam. Pp. 127–183.

Carluccio, A. 1989. *A Passion for Mushrooms*. Pavillion, London. Carter, Z. 1994. *Tuber melanosporum*. *North American Truffling Society Newsletter* 12(1): 2–3.

Castellano, M.A. 1995. NATS truffle and truffle-like fungi. 3. *Amogaster viridigleba* gen. et sp. nov. *Mycotaxon* 55:179–185.

Cave, G.L. 2005. Risk analysis for *Phytophthora ramorum* Werres, de Cock and In't Veld, causal agent of *Phytophthora* canker (sudden oak death), ramorum leaf blight, and ramorum dieback. Available via www.aphis.usda.gov/ppq/ispm/pramorum/pramorumpra05-05-05.pdf.

Cázares, E., Luoma, D.L., Amaranthus, M.P., Chambers, C.L., Lehmkuhl, J.F. 1999. Interaction of fungal sporocarp production with small mammal abundance and diet in Douglas-fir stands of the southern Cascade Range. *Northwest Science* 73:64–76.

Centers for Disease Control. 2006. Division of bacterial and mycotic diseases: food

irradiation. Available via www.cdc.gov/ncidod/dbmd/diseaseinfo/foodirradiation. htm#whatis.

Centre technologic forestall de Catalunya. 2001. Introduction of mycorrhizal edible fungi in reforestation sites of NE Spain, Universitat de Lleida. Available via www. labpatfor.udl.es/plantmicol/plantmicoleng.html.

Centro sperimentale di tartuficoltura. 2005. Tartuficoltura. Servizio Sistema Agroalimentare Ambiente Rurale e Foreste. Available via www.agri.marche.it/ Aree%20tematiche/Tartufi/TARTUFICOLTURA.htm.

Ceruti, A. 1960. Elaphomycetales et Tuberales. In: Bresadola, G., ed. *Iconographia micologica*, Vol. 28(Suppl. 2). Trento.

Ceruti, A. 1990. Evoluzione delle conoscenze biologiche sul genere *Tuber*. In: Bencivenga, M., Granetti, B., eds. *Atti del secondo congresso internazionale sul tartufo*. Spoleto, Italy, 24–27 November 1988. Comunità Montana dei Monti Martani e del Serano, Spoleto. Pp. 1–16.

Ceruti, A., Fontana, A., Nosenzo, C. 2003. *Le specie europee del genere* Tuber *una revisione storica*. Monografie no. 37. Regione Piemonte, Museo Regionale di Scienze Naturali, Torino.

Cetto, B. 1989–1992. *I funghi dal vero*. Vols. 1–7. Saturnia, Trento.

Chang, S.-T. 1993. Mushroom biology: the impact on mushroom production and mushroom products. In: Chang, S.-T., Buswell, J.A., Chiu, S.-W., eds. *Mushroom Biology and Mushroom Products*. Chinese University Press, Hong Kong. Pp. 3–20.

Chatin, G.A. 1892. *La truffe: botanique de la truffe et des plantes truffiaeres-sol-climat-pays producteurs-conserves-prâeparations culinaires*. Baillière, Paris.

Chen, A. 2001. Cultivation of *Lentinula edodes* on synthetic logs. *The Mushroom Growers' Newsletter*, August. Available via www.mushroomcompany.com/200108/ shiitake.pdf.

Chen, D.M., Taylor, A.F.S., Burke, R.M., Cairney, W.G. 2001. Identification of genes for lignin peroxidases and manganese peroxidases in ectomycorrhizal fungi. *New Phytologist* 152:151–158.

Chen, J., Liu, P.-G., Wang, Y. 2005. Notes on *Tuber aestivum* (Tuberaceae, Ascomycota) from China. *Acta Botanica Yunnanica* 27:385–389.

Chen, Y.L., Dell, B., Le Tacon, F. 2001. Chinese truffles (*Tuber*): diversity and their geographical distribution. Paper presented at the Third International Conference on Mycorrhizas, Adelaide, Australia. Available via www.mycorrhiza.ag.utk. edu/latest/icoms/icom3/ICOM3_chen-y.htm.

Chengdu Oriental Foodstuff Trading Co. Ltd. 2007. http://218.246.702.76:8080/ index.asp.

Chevalier, G. 1973. Synthèse axénique des mycorhizes de *Tuber brumale* Vitt. a partir de cultures pures du champignon. *Annales de phytopathologie* 2:163–182.

Chevalier, G. 1985. Synthèse des mycorhizes de différents *Tuber* avec *Corylus avellana* et *Quercus pubescens* sous aerosol nutritive. *Agronomie* 6:563–564.

Chevalier, G. 1998. The truffle cultivation in France: assessment of the situation after 25 years of intensive use of mycorrhizal seedlings. In: Danell, E., ed. *Proceedings of the First International Meeting on Ecology, Physiology, and Cultivation of Edible Myc-*

orrhizal Mushrooms. Uppsala, Sweden, 3–4 July 1998. Available via www.icom2.slu. se/ABSTRACTS/Bencivenga.html.

Chevalier, G. 2001. From the Spoleto congress to the one in Aix-en-Provence: projections regarding researches on truffle and truffle cultivation in France. In: *Proceedings of the Fifth International Congress on the Science and Cultivation of Truffles*. Aix-en-Provence, France, 3–6 March 1999. Federation Française des Trufficulteurs, Paris. Pp. 11–15.

Chevalier, G., Desmas, C. 1975. Synthèse axénique des mycorhizes de *Tuber melanosporum*, *T. uncinatum* et *T. rufum* sur *Pinus sylvestris* a partir de cultures pures du champignon. *Annales de phytopathologie* 7:338.

Chevalier, G., Desmas, C., Frochot, H., Riousset, L. 1978. L'espèce *Tuber aestivum* Vitt. I. Définition. *Mushroom Science* 10:957–975.

Chevalier, G., Dupré, C. 1990. Recherche et expérimentation sur la truffe et la trufficulture en France. In: Bencivenga, M., Granetti, B., eds. *Atti del secondo congresso internazionale sul tartufo*. Spoleto, Italy, 24–27 November 1988. Comunità Montana dei Monti Martani e del Serano, Spoleto. Pp. 157–166.

Chevalier, G., Frochot, H. 1997. *La truffe de Bourgogne* (*Tuber uncinatum* Chatin). Petraque, Levallois-Perret.

Chevalier, G., Gregori, G., Frochot, H., Zambonelli, A. 2002. The cultivation of the Burgundy truffle. In: Hall, I.R., Wang, Y., Danell, E., Zambonelli, A., eds. *Edible Mycorrhizal Mushrooms and Their Cultivation: Proceedings of the Second International Conference on Edible Mycorrhizal Mushrooms*. Christchurch, New Zealand, 3–5 July 2001. CD-ROM. New Zealand Institute for Crop and Food Research Limited, Christchurch.

Chevalier, G., Grente, J. 1973. Propagation de la mycorhization par la truffe à partir de racines excisées et de plantes inséminateurs. *Annales de phytopathologie* 5:317–318.

Chevalier, G., Grente, J. 1978. Application pratique de la symbiose ectomycorhizienne: Production a grande échelle de plantes mycorhizes par la truffe (*Tuber melanosporum* Vitt.). *Mushroom Science* 10:483–505.

Chevalier, G., Mousain, D., Couteaudier, Y. 1975. Association ectomicorrhiziennes entre des Tuberácéae et des Cistacéae. *Annals de phytopathologia* 7:355–356.

Chevalier, G., Poitou, N. 1990. Facteurs conditionnant l'utilisation optimale des plants mycorhizés artificiellement par la truffe. In: Bencivenga, M., Granetti, B., eds. *Atti del secondo congresso internazionale sul tartufo*. Spoleto, Italy, 24–27 November 1988. Comunità Montana dei Monti Martani e del Serano, Spoleto. Pp. 409–413.

Chevalier, G., Riousset, L., Riousset, G., Dupré, C. 1990. Taxonomie des truffes européennes. In: Bencivenga, M., Granetti, B., eds. *Atti del secondo congresso internazionale sul tartufo*. Spoleto, Italy, 24–27 November 1988. Comunità Montana dei Monti Martani e del Serano, Spoleto. Pp. 37–44.

Chin, G., Mommaneni, S. 2005. The silk industry in Japan in the 1800s. Available via www.smith.edu/hsc/silk/papers/chin.html.

Chinese.truffle.com. 2005. www.chinesetruffle.com/chinesetruffle.html.

Ciampolini, M., Suss, L. 1982–1983. Nuovi reperti sulla mosca dell'aglio, *Suilla univittata* (von Roser) (Diptera Heleomyzidae). *Bollettino di zoologia agraria e di bachicoltura* 17:19–38.

Ciani, A. 1990. Il circuito commerciale del tartufo in Italia. In: Bencivenga, M., Granetti, B., eds. *Atti del secondo congresso internazionale sul tartufo.* Spoleto, Italy, 24–27 November 1988. Comunità Montana dei Monti Martani e del Serano, Spoleto. Pp. 621–631.

Ciani, A., Granetti, B., Vincenti, D. 1992. Il tartufo in Italia e nel mondo: aree di produzione, mercato e prezzi. *L'informatore agrario* 47:51–62.

Ciani, A., Sciarresi, C., Martino, G., Ricci, F. 1988. Tartuficoltura e recupero delle aree interne. *L'informatore agrario* 44:209–221.

Ciccarelli, A. 1564. *Opusculum de tuberibus.* Pavia.

Citterio, B., Cardoni, P., Potenza, L., Amicucci, A., Gola, G., Nuti, M.P. 1995. Isolation of bacteria from sporocarps of *Tuber magnatum* Pico, *Tuber borchii* Vitt. and *Tuber maculatum* Vitt.: identification and characterization. In: Stocchi, V., Bonfante, P., Nuti, M., eds. *Biotechnology of Ectomycorrhizae: Molecular Approaches.* Plenum, New York. Pp. 241–248.

Claridge, A.W., Castellano, M.A., Trappe, J.M. 1996. Fungi as a food resource for mammals in Australia. *Fungi of Australia.* Vol. 1B. CSIRO, Collingwood, Australia. Pp. 239–267.

Claus, R., Hoppen, H.O., Karg, H. 1981. The secret of truffles: A steroidal pheromone? *Experimentia* 37:1178–1179.

Clavel, G. 2005. La truffe, "diamant noir" des gourmets, mais attention aux imitations. *ModeMag* 51, 20 December.

Clavel, G. 2007. Buyer beware: truffle market full of fakes. Available via www.expatica.com/actual/article.asp?subchannel_id=209&story_id=26533.

Colgan III, W., Carey, A.B., Trappe, J.M., Molina, R.J., Thysell, D. 1999. Diversity and productivity of hypogeous fungal sporocarps in a variably thinned Douglas-fir forest. *Canadian Journal of Forest Research* 29:1259–1268.

Colgan III, W., Trappe, J. 1997. NATS truffle and truffle-like fungi. 7. *Tuber anniae* sp. nov. (Ascomycotina). *Mycotaxon* 64:437–441.

Colinas, C., Bonet, J.A., Fischer, C. 2001. Truffle cultivation: an alternative to agricultural subsidies. In: *Proceedings of the Fifth International Congress on the Science and Cultivation of Truffles.* Aix-en-Provence, France, 3–6 March 1999. Federation Française des Trufficulteurs, Paris.

Comandini, O., Pacioni, G. 1997. Mycorrhizae of Asian black truffles, *Tuber himalayense* and *T. indicum. Mycotaxon* 63:77–86.

Commonwealth Scientific and Industrial Research Organization. 2001. Plant roots in drains: prevention and cure. CSIRO Building Technology File no. 17. Available via www.publish.csiro.au/samples/BTF17Sample.pdf.

Consiglio regionale della regione del Veneto. 1988. Disciplina della raccolta, coltivazione e commercializzazione dei tartufi. Legge regionale 28 giugno 1988, no. 30 (BUR no. 40/1988): Testo storico. L.R. 17/1988. Available via www.consiglioveneto.it/crvportal/leggi_storico/1988/88lr0030.html.

Consiglio regionale del Piemonte. 2006. Leggi regionali. Available via www.arianna.consiglioregionale.piemonte.it/.

Cooke, M.C. 1898. *A Plain and Easy Account of British Fungi.* Allen, London. Available via www.freespace.virgin.net/mikea.walton/#fungus.

Cooke, M.C., Massee, G. 1892. Himalayan truffles. *Grevillea* 20:67.

Cooper, K.M. 1976. A field survey of mycorrhizas in New Zealand ferns. *New Zealand Journal of Botany* 14:169–181.

Cornforth, I.S., Sinclair, A.G. 1984. *Fertiliser and Liming Recommendations for Pastures and Crops in New Zealand.* New Zealand Ministry of Agriculture and Fisheries, Wellington.

Coughan, A.P., Piché, Y. 2003. Method for synthesizing ectomycorrhiza *in vitro.* U.S. patent application no. 60/459,993.

Coughan, A.P., Piché, Y. 2005. *Cistus incanus* root organ cultures: a valuable tool for studying mycorrhizal associations. In: Declerck, S., Strullu, D.G., Fortin, A., eds. *In Vitro Culture of Mycorrhizas.* Springer, Heidelberg. Pp. 235–249.

Courtecuisse, R., Duhem, B. 1995. *Mushrooms and Toadstools of Britain and Europe.* Collins, London.

Courvoisier, M. 1995. France: Les importations de truffes fraîches en provenance de Chine. *Le trufficulteur français* 11:10–11.

Craddock, J.H. 1994. Mycorrhizal association between *Corylus heterophylla* and *Tuber melanosporum. Acta Horticulturae* 351:291–298.

Cranshaw, W.S. 2006. *Bacillus thuringiensis.* Colorado State University Cooperative Extension Report no. 5.556. Available via www.ext.colostate.edu/PUBS/INSECT/05556.html.

Dahlberg, A. 2001. Community ecology of ectomycorrhizal fungi: an advancing interdisciplinary field. *New Phytologist* 150:555–562.

Danell, E., ed. 1998. *Proceedings of the First International Meeting on the Ecology, Physiology and Culture of Edible Mycorrhizal Mushrooms.* Uppsala, Sweden, 3–4 July 1998.

Danell, E. 2000a. *Cantharellus.* In: Cairney, J.W.G., Chambers, S.M., eds. *Ectomycorrhizal Fungi: Key Genera in Profile.* Springer Verlag, Berlin. Pp. 253–267.

Danell, E. 2000b. Cultivation of edible ectomycorrhizal mushrooms: state of the art. In: Fortin, J.A., Piche, Y., eds. *Les Champignons forestiers: recolte, commercialisation et conservation de la resource.* CRBT, Universite Laval, Quebec. Pp. 17–20.

Danielson, R.M. 1984. Ectomycorrhiza formation by the operculate discomycete *Sphaerosporella brunnea* (Pezizales). *Mycologia* 76(3):454–461.

Datta, R., Nanavaty, M. 2005. *Global Silk Industry: A Complete Source Book.* Universal Publishers, Boca Raton, Fla.

David, E. 1981. *French Provincial Cooking.* Penguin, Middlesex.

Davisnet. 2005. www.davisnet.com/weather/cool/world.asp.

Deacon, J. 2006. The microbial world: *Bacillus thuringiensis.* Available via www.helios.bto.ed.ac.uk/bto/microbes/bt.htm.

De Bary, H.A. 1879. *Die Erscheinung der Symbiose.* Strassbourg.

Delatour, C., Saurat, C., Husson, C., Ioos, R., Schenck, N. 2002. Discovery of *Phytophthora ramorum* on *Rhododendron* sp. in France and experimental symptoms on *Quercus robur.* In: *Proceedings of the Sudden Oak Death Symposium: The State of Our Knowledge.* Monterey, California, 15–18 December 2002. Available via www.danr.ucop.edu/ihrmp/sodsymp/poster/poster57.html.

De Lisle, R. 2003. A whiff of white gold. *Daily Telegraph*, 15 November. Available via www.telegraph.co.uk/wine/main.jhtml?xml=/wine/2003/11/15/edtruf15.xml.

Dell, B., Malajczuk, N., Bougher, N.L., Thomson, G. 1994. Development and function of *Pisolithus* and *Scleroderma* ectomycorrhizas formed *in vivo* with *Allocasuarina*, *Casuarina*, and *Eucalyptus*. *Mycorrhiza* 5:129–138.

Delmas, J. 1978. *Tuber* spp. In: Chang, S.T., Hayes, W.A., eds. *The Biology and Cultivation of Edible Mushrooms*. Academic Press, London. Pp. 645–681.

Delmas, J. 1983. *La Truffe et sa culture*. Institut National de la Recherche Agronomique, Paris.

Delmas, J., Poitou, N. 1978. La mycorhization de *Quercus pubescens* par *Tuber melanosporum* en conditions contrôlées: Influence de quelques facteurs du milieu. *Mushroom Science* 10:995–1005.

Delmas, M., Gaset, A., Montant, C., Pébeyre, P.-J., Talou, T. 1990. Process for the production of an aromatic product having the odor and taste of black truffles, product and aromatic body obtained. U.S. patent 4,906,487.

Den Bakker, H.C., Zuccarello, G.C., Kuyper, T.W., Noordeloos, M.E. 2004. Evolution and host specificity in the ectomycorrhizal genus *Leccinum*. *New Phytologist* 163:201–215.

Denman, S., Kirk, S.A., Brasier, C.M. 2005. *Phytophthora ramorum* on *Quercus ilex* in the United Kingdom. *Plant Disease Note* 89:1241. Available via www.apsnet.org/pd/searchnotes/2005/pd-89-1241a.asp.

Department for Environment, Food, and Rural Affairs. 2003. England Rural Development Programme. Appendix A6: East of England region. Section 1: Description of current situation in the east of England region. Available via www.defra.gov.uk/erdp/pdfs/programme/east/section1_1pages7to12.pdf.

Department for Environment, Food and Rural Affairs. 2005. CSL pest risk analysis for a new *Phytophthora* species informally named as *Phytophthora kernoviae* (also known as *P. kernovii*; formally *Phytophthora* taxon C). Available via www.defra.gov.uk/planth/pra/forest.pdf.

Department for Environment, Food, and Rural Affairs. 2005. Passporting guide. Available via www.defra.gov.uk/planth/publicat/passport/pass.pdf.

De Roman, M., Claveria, V., De Miguel, A.-M. 2005. A revision of the descriptions of ectomycorrhizas published since 1961. *Mycological Research* 109:1063–1104.

De Simone, C., Lorenzoni, P., Raglione, M. 1993. Il manganese nei suoli di produzione di *Tuber melanosporum* e di *Tuber aestivum*. *Annali Facoltà di Agraria Università degli Studi di Sassari* (I), 35(2):415–418.

Dexheimer, J., Gerard, J., Leduc, J.P., Chevalier, G. 1985. Etude ultrastructurale comparée des associations symbiotiques mycorhiziennes *Helianthemum salicifolium–Terfezia claveryi* et *Helianthemum salicifolium–Terfezia leptoderma*. *Canadian Journal of Botany* 63:582–591.

Díaz, P., Ibañez, E., Señoráns, F.J., Reglero, G. 2003. Truffle aroma characterization by headspace solid-phase microextraction. *Journal of Chromatography A* 1017:207–214.

Díaz, P., Señoráns, F.J., Reglero, G., Ibañez, E. 2002. Truffle aroma analysis by

headspace solid phase microextraction. *Journal of Agricultural Food Chemistry* 50:6468–6472.

Di Massimo, G., Bencivenga, M., Tedeschini, E., Garcia Montero, G., Manjon, J.L. 1998. Nuova specie di *Tuber* importata dall'oriente. *Micologia Italiana* 27(1):13–18.

Dobermann, A., Fairhurst, T. 2000. *Rice: Nutrient Disorders and Nutrient Management*. Potash and Phosphate Institute (PPI), Potash and Phosphate Institute of Canada (PPIC), and International Rice Research Institute. Available via www.knowledgebank.irri.org/riceDoctor_MX/Fact_Sheets/DeficienciesToxicities/Manganese_Deficiency.htm.

Donnini, D. 2006. Truffle cultivation in Italy: results and perspectives. In: *Abstracts of the International Truffle Orchards Workshop*. Juva, Finland, 16–18 October. Pp. 16–17.

Donnini, D., Baciarelli Falini, L., Bencivenga, M. 2001. Analisi della micorrizazione in tartufaie coltivate di *Tuber melanosporum* Vittad. impiantate da oltre 12 anni in ambienti pedoclimatici diversi. In: *Proceedings of the Fifth International Congress on the Science and Cultivation of Truffles*. Aix-en-Provence, France, 3–6 March 1999. Federation Française des Trufficulteurs, Paris. Pp. 437–440.

Doumenc-Faure, M., Giacinti-Martine, G., Talou, T. 2001. L'arome de la truffe noire (*Tuber melanosporum* Vitt.): de l'etude de l'effect de sol a l'authentification d'aromatisants. In: *Proceedings of the Fifth International Congress on the Science and Cultivation of Truffles*. Aix-en-Provence, France, 3–6 March 1999. Federation Française des Trufficulteurs, Paris. Pp. 142–146.

Downer, J., Faber, B., Menge, J. 2002. Factors affecting root rot control in mulched avocado orchards. *Hort Technolology* 12:601–605.

Dubé, S.L. 2003. Vitro truffle mycorrhized hazels and oaks for the establishment of truffle orchards are unrestrained travellers. In: *Proceedings of the Third International Meeting on the Ecology, Physiology and Culture of Edible Mycorrhizal Mushrooms*. Victoria, 15–22 August 2003. 2 CD-ROMs. University of Victoria, Victoria, B.C.

Ducret, J.P. 2001. Radiesthesie et trufficulture. In: *Proceedings of the Fifth International Congress on the Science and Cultivation of Truffles*. Aix-en-Provence, France, 3–7 March 1999. Paris, Federation Français des trufficulteurs. Pp. 205–207.

Dunstan, W.A., Dell, B., Malajczuk, N. 1998. The diversity of ectomycorrhizal fungi associated with introduced *Pinus* spp. in the Southern Hemisphere, with particular reference to Western Australia. *Mycorrhiza* 8:71–79.

Edinformatics. 2005. Truffle. Available via www.edinformatics.com/culinaryarts/food_encyclopedia/truffle.htm.

Edwards, B. 1999. Correcting manganese deficiencies in eastern North Carolina. In: Allen, J.L., ed. *Proceedings of the 42nd Annual Meeting of the Soil Science Society of North Carolina*. Raleigh, North Carolina, 19–20 January 1999. Soil Science Society of North Carolina, Raleigh. Pp. 35–39. Available via www.ncagr.com/agronomi/MNSSSNC.htm.

Elder, R.J., Reid, D.J., Macleod, W.N.B., Gillespie, R.L. 2002. Post-ratoon growth and yield of three hybrid papayas (*Carica papaya* L.) under mulched and bare-ground conditions. *Australian Journal of Experimental Agriculture* 42:71–81.

Ellis, R.J., Bragdon, G.A., Schlosser, B.J. 1999. Properties of the blue light requirements for primordia initiation and basidiocarp maturation in *Coprinus stercorarius*. *Mycological Research* 103:779–784.

Ericsson, L. 2001. Cultivation of the Burgundy truffle in Sweden. MSc. thesis, Swedish University of Agricultural Sciences, Uppsala. Available via www.mykopat.slu. se/newwebsite/ex/lina/lina.html.

Eriksson, O.E., Baral, H.O., Currah, R.S., Hansen, K., Kurtzman, C.P., Laessøe, T., Rambold, G. 2002. Myconet: notes on Ascomycetes systematics. Available via www.fieldmuseum.org/myconet/printed_v8.asp.

Etayo, M.L., De Miguel, A.M. 2001. Effect of mulching on *Tuber melanosporum* Vitt. mycorrhizae in a cultivated truffle bed vs. other competing mycorrhizae. In: *Proceedings of the Fifth International Congress on the Science and Cultivation of Truffles*. Aix-en-Provence, France, 3–6 March 1999. Federation Française des Trufficulteurs, Paris. Pp. 378–381.

Eyre, W.L.W. 1900. Fungi. In: Doubleday, H.A., ed. *Victoria County History of Hampshire and the Isle of Wight*. Vol. 1. University of London, London. Pp. 82–87.

Fédération de la Interrégionale des Trufficulteurs du Centre et de l'Est. 2007. La truffe de Bourgogne. Available via www.truffe-de-bourgogne.com/especes.htm.

Feeney, J. 2002. Desert truffles galore. *Saudi Aramco World* 53(5). Available via www. saudiaramcoworld.com/issue/200205/desert.truffles.galore.htm.

Ferdman, Y., Aviram, S., Roth-Bejerano, N., Trappe, J.M., Kagan-Zur, V. 2005. Phylogenetic studies of *Terfezia pfeilii* and *Choiromyces echinulatus* (Pezizales) support new genera for southern African truffles: *Kalaharituber* and *Eremiomyces*. *Mycological Research* 109:237–245.

Ferrari, M., Menta, A., Marcon, E., Montetermini, A. 1999. *Malattie e parassiti delle piante da fiore ornamentali e forestali*. Edagricole, Bologna.

Fioc, L. 1987. *La truffe telle que je la pratique*. Fioc, Saint Paul-Trois-Chateaux.

Fischer, C., Colinas, C. 1996a. Fase 3. Metodología y resultados de la síntesis de micorrizas entre *Quercus ilex* y 3 especies de hongos del género *Tuber*. In *Metodo de control de planta de Quercus ilex inoculada con Tuber melanosporum*. Informe a la Junta de Castilla y León.

Fischer, C., Colinas, C. 1996b. Methodology for certification of *Quercus ilex* seedlings inoculated with *Tuber melanosporum* for commercial application. Poster presented at the First International Conference on Mycorrhizae. Berkeley, California, August 1996. Available via http//:labpatfor.udl.es/docs/tubing.html.

Fischer, C., Colinas, C. 2005. Germination of black truffle ascospores. In: *Proceedings of the Fourth International Workshop on Edible Mycorrhizal Mushrooms*. Murcia, Spain, 28 November–2 December 2005. Universidad Murcia, Murcia. P. 49.

Flück, M. 1995. *Welcher pilz ist das?* Franckh-Kosmos Verlag, Stuttgart.

Fontana, A. 1967. Sintesi micorrizica tra *Pinus strobus* e *Tuber maculatum*. *Giornale Botanico Italiano* 101:298–299.

Fontana, A., Bonfante Fasolo, P. 1971. Sintesi micorrizica di *Tuber brumale* Vitt. con *Pinus nigra* Arnold. *Allionia* 17:15–18.

Fontana, A., Palenzona, M. 1969. Sintesi micorrizica di *Tuber albidum* in coltura pura con *Pinus strobus* e pioppo euroamericano. *Allionia* 15:99–104.

Food and Fertilizer Technology Center. 2001. Manganese deficiency of crops: soybean (*Glycine max* L.). Available via www.agnet.org/library/bc/51004/.

Fortas, Z., Chevalier, G. 1992. Effet des conditions de culture sur la mycorhization de l'*Helianthemum guttatum* par trois espèces de terfez des genres *Terfezia* et *Tirmania* d'Algérie. *Canadian Journal of Botany* 70:2453–2460.

Fortin, A., Bécard, G., Declerck, S., Dalpé, Y., St-Arnaud, M., Coughlan, A.P., Piché, Y. 2002. Arbuscular mycorrhiza on root-organ cultures. *Canadian Journal of Botany* 80:1–20.

Foundation for Research Science and Technology. 2005. Briefing to the incoming minister. www.frst.govt.nz/publications/corporate/downloads/BIM/20051018_Briefing_to_Incoming_Minister.pdf.

Founoune, H., Duponnois, R., Ba, A.M., Sall, S., Branget, I., Lorquin, J., Neyra, M., Chotte, J.L. 2002. Mycorrhiza helper bacteria stimulated ectomycorrhizal symbiosis of *Acacia holosericea* with *Pisolithus alba*. *New Phytologist* 153:81–89.

Francolini, F. 1931. *Tartuficoltura e rimboschimenti*. Federazione Italiana dei Consorzi Agrari, Piacenza.

Frank, A.B. 1877. Über die biologischen Verhältnisse des Thallus einiger Krustenflechten. *Beiträge zur Biologie der Pflanzen* 2:123–200.

Frank, A.B. 1888. Über die physiologische Bedeutung der Mycorhiza. *Berichte der Deutschen Botanischen Gesellschaft* 6:248–269.

Frochot, H., Chevalier, G., Bardet, M.C., Aubin, J.P. 1990. Effet de la désinfection du sol et des antécédents culturaux sur l'évolution de la mycorhization avec *Tuber melanosporum* sur noisetier. In: Bencivenga, M., Granetti, B., eds. *Atti del secondo congresso internazionale sul tartufo*. Spoleto, Italy, 24–27 November 1988. Comunità Montana dei Monti Martani e del Serano, Spoleto. Pp. 289–296.

Gadoury, D. 2002. Biological control: a guide to natural enemies in North America—*Ampelomyces quisqualis* (Deuteromycetes). Available via www.nysaes.cornell.edu/ent/biocontrol/pathogens/ampelomyces.html.

Galbraith, A. 1988. Truffle pursuit. *Sunday Express Magazine* (London), 4 December, 27–29.

Gale, G. 2003. Saving the vine from *Phylloxera*: a never-ending battle. In: Sandler, M., Pinder, R., eds. *Wine: A Scientific Exploration*. Taylor and Francis, London. Pp. 70–91. Available via www.cas.umkc.edu/philosophy/gale/proofs.pdf.

Gandeboueuf, D., Drupe, C., Henrion, B., Martin, F., Chevalier, G. 1996. Characterization and identification of *Tuber* species using biochemical and molecular criteria. In: Azcon-Aguilar, C., Barea, J.M., eds. *Mycorrhizas in Integrated System from Genes to Plant Development: Proceedings of the Fourth European Symposium on Mycorrhizae*. Granada, Spain, 11–14 July 1994. European Community, Luxenbourg. Pp. 31–34.

Garbaye, J. 1994. Helper bacteria: a new dimension to the mycorrhizal symbiosis. Tansley review no. 76. *New Phytologist* 128:197–210.

Garbaye, J., Churin, J.-L., Duponnois, R. 1992. Effects of substrate sterilization, fungicide treatment, and mycorrhization helper bacteria on ectomycorrhizal formation of pedunculate oak (*Quercus robur*) inoculated with *Laccaria laccata* in two peat bare-root nurseries. *Biology and Fertility of Soils* 13:55–57.

Gardin, L. 2005. *I tartufi minori in Toscana*. Arsia, Florence.

Garland, F. 1996. *Truffle Cultivation in North America*. Garland Mushrooms and Truffles Inc., Hillsborough, N.C.

Garland, F. 2001. Growing *Tuber melanosporum* under adverse acid soil conditions in the United States of America. In: *Proceedings of the Fifth International Congress on the Science and Cultivation of Truffles*. Aix-en-Provence, France, 3–6 March 1999. Federation Française des Trufficulteurs, Paris.

Garvey, D.C., Cooper, P.B. 2001. *French Black Truffle Establishment and Production in Tasmania*. Rural Industries Research and Development Corporation, Australia, Publication no. 01/084.

Gateway Africa. 2006. www.gateway-africa.com/fuanaflora/Plants/acacia_hebeclada.html.

Gazzetta ufficiale della Republica Italiana. 2006. Attuazione della direttiva 1999/105/CE relativa alla commercializzazione dei materiali forestali di moltiplicazione. Available via www.camera.it/parlam/leggi/deleghe/testi/03386dl.htm.

Genbank. 2004. www.ncbi.nlm.nih.gov/Genbank/GenbankOverview.html.

Genc, C. 2006. World fertilizer use manual: hazelnut or filbert (*Corylus avellana* L.). International Fertilizer Industry Association. Available via www.fertilizer.org/ifa/publicat/html/pubman/hazelnut.htm.

Georgia Faces. 2002. Georgia orchards hide pecan truffle bonuses. Available via www.georgiafaces.caes.uga.edu/getstory.cfm?storyid=1732.

Gerard, J. 1597. *Herball, Generall Historie of Plants*. Reprint 1985. Crescent, New York.

Gibelli, G. 1883. Nuovi studi sulla malattia del castagno detta dell'inchiostro. *Memorie dell' Accademia delle Scienze dell'Istituto di Bologna* 4:287–314.

Gioacchini, A.M., Menotta, M., Bertini, L., Rossi, I., Zeppa, S., Zambonelli, A., Piccoli, G., Stocchi, V. 2005. Solid-phase microextraction gas chromatography/mass spectrometry: a new method for species identification of truffles. *Rapid Communications in Mass Spectrometry* 19:2365–2370.

Giomaro, G., Sisti, D., Zambonelli, A. 2005. Cultivation of edible ectomycorrhizal fungi by *in vitro* mycorrhizal synthesis. In: Declerck, S., Strullu, D.-G., Fortin, J.A., eds. *In Vitro Cultivation of Mycorrhizas*. Springer, Berlin. Pp. 253–270.

Giomaro, G., Sisti, D., Zambonelli, A., Amicucci, A., Cecchini, M., Comandini, O., Stocchi, V. 2002. Comparative study and molecular characterization of ectomycorrhizas in *Tilia americana* and *Quercus pubescens* with *Tuber brumale*. *FEMS Microbiology Letters* 216:9–14.

Giovannetti, G. 1980. Method of producing plants mycorryzated with symbiotic fungi. Italian patent 1,128,367; U.S. patent 4,345,403 (1982).

Giovannetti, G. 1990. Prima produzione di carpofori di *Tuber magnatum* Pico da piante micorizate fornite da vivai specializzati. In: Bencivenga, M., Granetti, B., eds. *Atti del secondo congresso internazionale sul tartufo*. Spoleto, Italy, 24–27 November 1988. Comunità Montana dei Monti Martani e del Serano, Spoleto. Pp. 297–302.

Giovannetti, G., Fontana, A. 1980–1981. Simbiosi micorrizica di *Tuber macrosporum* Vitt. con alcune Fagales. *Allionia* 24:13–17.

Giovannetti, G., Roth-Bejerano, N., Zanini, E., Kagan-Zur, V. 1994. Truffles and their cultivation. *Horticultural Reviews* 16:71–107.

Girard, M., Verlhac, A. 1987. Experimentation truffière. *Infos* 32:23–28.

Glamočlija, J., Vujičić, R., Vukojević, J. 1997. Evidence of truffles in Serbia. *Mycotaxon* 65:211–222.

Goldway, M., Amir, R., Goldberg, D., Hadar, Y., Levanon, D. 2000. *Morchella conica* exhibiting a long fruiting season. *Mycological Research* 104:1000–1004.

Gomez Fernandez, J., Moreno Arroyo, B. 1995. Contribución al conocimiento del género *Tuber* (Micheli ex Wiggers: Fr.) el la provincia de Jaen. I. *Lactarius* 4:40–46.

Govi, G., Bencivenga, M., Granetti, B., Pacioni, G., Palenzona, M., Tocci, A., Zambonelli, A. 1997. Metodo basato sulla caratterizzazione morfologica delle micorrize. In: *Regione Toscana: Il tartufo*. Compagnia delle Foreste, Arezzo. Pp. 148–155.

Graebner, L. 1991. Despite solid technology, truffle firm files for bankruptcy. *The Business Journal* (Sacramento) 8(4):9.

Granetti, B. 1995. Caratteristiche morfologiche, biometriche e strutturali delle micorrize di *Tuber* di interesse economico. *Micologia Italiana* 24(2):101–117.

Granetti, B., De Angelis, A., Materazzi, G. 2005. *Umbria terra di tartufi, Regione Umbria*. Ubriagraf, Terni.

Granetti, B., Minigrucci, G., Bricchi, E. 1990. Analisi biometrica e morfologica delle ascospore di alcune specie del genere *Tuber*. In: Bencivenga, M., Granetti, B., eds. *Atti del secondo congresso internazionale sul tartufo*. Spoleto, Italy, 24–27 November 1988. Comunità Montana dei Monti Martani e del Serano, Spoleto. Pp. 59–100.

Gregori, G. 1991. *Tartufi e tartuficoltura nel Veneto*. Regione del Veneto, Assessorato Agricoltura e Foreste, Dipartimento Foreste, Padova. Tipografia, Vicenza.

Gregori, G. 2002. Problems and expectations with the cultivation of *Tuber magnatum*. In: Hall, I.R., Wang, Y., Danell, E., Zambonelli, A., eds. *Edible Mycorrhizal Mushrooms and Their Cultivation: Proceedings of the Second International Conference on Edible Mycorrhizal Mushrooms*. Christchurch, New Zealand, 3–5 July 2001, CD-ROM. Christchurch, New Zealand Institute for Crop and Food Research Limited.

Gregori, G., Cecchini, M., Elisei, S., Pasqualini, L., Sacchi, A., Spezi, D. 2001. Tartufaie controllate di *T. magnatum* Pico: prove di miglioramento. In: *Proceedings of the Fifth International Congress on the Science and Cultivation of Truffles*. Aix-en-Provence, France, 3–6 March 1999. Federation Française des Trufficulteurs, Paris. Pp. 394–399.

Gregori, G., Ciapelloni, R. 1990. Produzione di piantine micorrize con *T. magnatum* Pico. In: Bencivenga, M., Granetti, B., eds. *Atti del secondo congresso internazionale sul tartufo*. Spoleto, Italy, 24–27 November 1988. Comunità Montana dei Monti Martani e del Serano, Spoleto. Pp. 211–218.

Gregori, G., Tocci, A. 1985. Possibilità di produzione di piantine di *Alnus cordata* Loisel micorrizate con *T. melanosporum* Vitt. *Tuber aestivum* Vitt. *L'Italia forestale e montana* 40(5):262–270.

Grente, J. 1972–1974. *Perspectives pour une trufficulture moderne*. INRA, Clermont-Ferrand.

Grente, J., Chevalier, G., Pollacsek, A. 1972. La germination de l'ascospore de *Tuber melanosporum* et la synthèse sporale des mycorhizes. *Comptes rendus hebdomadaires des seances de l'Academie des sciences, Serie D* 275:743–746.

Grente, J., Delmas, J., Poitou, N., Chevalier, G. 1974. Faits nouveaux sur la truffe. *Mushroom Science* 9:815–846.

Greuter, W., McNeill, J., Barrie, F.R., Burdet, H.-M., Demoulin, V., Filgueiras, T.S., Nicolson, D.H., Silva, P.C., Skog, J.E., Trehane, P., Turland, N.J., Hawksworth, D.L. 2000. International code of botanical nomenclature: electronic version. Available via www.bgbm.org/iapt/nomenclature/code/SaintLouis/0000St. Luistitle.htm.

Griffiths, E. 1978. *Soils of the Waikari District, North Canterbury, New Zealand*. New Zealand Soil Survey Report no. 29.

Groupement Européen *Tuber* (Grupo Europeo *Tuber*, Gruppo Europeo *Tuber*). 2002. Programme de developpement de la trufficulture et de reconstitution d'un verger truffier en Europe (2003–2007). Fédération Française des Trufficulteurs, Paris.

Gutierrez, A., Honrubia, M., Morte, A., Diaz, G. 1996. Edible fungi adapted to arid and semi-arid areas: molecular characterization and in vitro mycorrhization of micropropagated plantlets. In: *La mycorhization des plantes forestières en milieu aride et semi-aride et la lutte contre la désertification dans le bassin méditerranéen (Mycorrhization of Forest Plants under Arid and Semi-Arid Conditions and Desertification Control in the Mediterranean)*. Séminaire du groupe de travail CIHEAM sur l'utilisation des mycorhizes pour la lutte contre la désertification dans le Bassin Méditerranéen (MYCOLUDESME). Zaragoza, Spain, 15–17 November 1995. *Cahiers Options Méditerranéennes* 20:139–144. Available via www.ressources.ciheam.org/om/pdf/c20/96605784.pdf.

Hall, I.R. 1973. Endogonaceous fungi associated with rata and kamahi. Ph.D. diss., University of Otago, Dunedin, New Zealand.

Hall, I.R. 1976. Response of *Coprosma robusta* to different forms of endomycorrhizal inoculum. *Transactions of the British Mycological Society* 67:409–411.

Hall, I.R. 1988. Potential for exploiting vesicular-arbuscular mycorrhizas in agriculture. In: A. Mizrahi, ed. *Advances in Biotechnological Processes*. Vol. 9, *Biotechnology in Agriculture*. ARL, New York. Pp. 141–174.

Hall, I.R. 2006. Truffles & Mushrooms web site. www.trufflesandmushrooms.co.nz.

Hall, I.R., Brown, G., Byars, J. 2001. *The Black Truffle: Its History, Uses and Cultivation*. Reprint of 2nd ed. on CD-ROM plus booklet. New Zealand Institute for Crop and Food Research Ltd., Christchurch.

Hall, I.R., Dixon, C.A., Parmenter, G.A., Martin, N., Hance-Halloy, M.-L. 2002a. Factors affecting fruiting of the Périgord black truffle: a comparison of productive and non-productive *Tuber melanosporum* truffières in New Zealand. Crop and Food Research Confidential Report no. 768 on CD-ROM (restricted to members of the New Zealand Truffle Association). New Zealand Institute for Crop and Food Research Limited, Christchurch.

Hall, I.R., Stephenson, S.L., Buchanan, P.K., Wang, Y., Cole, A.L.J. 2003a. *Edible and Poisonous Mushrooms of the World*. Timber Press, Portland, Ore.

Hall, I.R., Wang, Y., Amicucci, A. 2003b. Cultivation of edible ectomycorrhizal mushrooms. *Trends in Biotechnology* 21:433–438.

Hall, I.R., Wang, Y., Danell, E., Zambonelli, A., eds. 2002b. *Edible Mycorrhizal Mushrooms and Their Cultivation: Proceedings of the Second International Conference on Edible Mycorrhizal Mushrooms*. CD-ROM. Christchurch, New Zealand Institute for Crop and Food Research Limited.

Hall, I.R., Zambonelli, A., Primavera, F. 1998. Ectomycorrhizal fungi with edible fruiting bodies. 3. *Tuber magnatum*, Tuberaceae. *Economic Botany* 52:192–200.

Hall, I.R., Zambonelli, A., Wang, Y. 2005. The cultivation of mycorrhizal mushrooms: success and failure. Paper presented at the Fifth International Conference on Mushroom Biology and Mushroom Products. *Acta Edulis Fungi* 12:45–60.

Hansen, J. 1982. A fungus in every pot. *New Scientist* 95:550–551.

Harkness, H.W. 1899. California hypogaeous fungi. *Proceedings of the California Academy of Sciences*, Ser. 3, 1:241–292.

Harley, J.H., Harley, E.L. 1987. A checklist of mycorrhiza in the British flora. *New Phytologist* (Suppl.) 105:1–102.

Harley, J.L., Smith, S.E. 1983. *Mycorrhizal Symbiosis*. Academic Press, London.

Hartley, M.J., Reid, J.B., Rahman, A., Springett, J.A. 1996. Effect of organic mulches and a residual herbicide on soil bioactivity in an apple orchard. *New Zealand Journal of Crop and Horticultural Science* 24:183–190.

Harvest Electronics. 2005. harvest.com/w.cgi?cmd=gph&hsn=3001&typ=4.

He, X.-Y., Li, H.-M., Wang, Y. 2004. *Tuber zhangdianense* sp. nov. from China. *Mycotaxon* 90:213–216.

Healy, R.A. 2003. *Mattirolomyces tiffanyae*, a new truffle from Iowa, with ultrastructural evidence for its classification in the Pezizaceae. *Mycologia* 95:765–772.

Heimsch, C. 1958. The first recorded truffle from Texas. *Mycologia* 50:657–660.

Henrion, B., Chevalier, G., Martin, F. 1994. Typing truffle species by PCR amplification of the ribosomal DNA spacers. *Mycological Research* 98(1):37–43.

Högberg, P., Plamboeck, A.H., Taylor, A.F.S., Fransson, P.M.A. 1999. Natural 13C abundance reveals trophic status of fungi and host-origin of carbon in mycorrhizal fungi in mixed forests. *Proceedings of the National Academy of Sciences USA* 96:8534–8539.

Hong Kong Observatory. 2007. www.hko.gov.hk/wxinfo/climat/world/eng/europe/europe_e.htm.

Honrubia, M., Gutiérrez, A., Morte, A. 2002. Desert truffle plantation from southeast Spain. In: Hall, I.R., Wang, Y., Danell, E., Zambonelli, A., eds. *Edible Mycorrhizal Mushrooms and Their Cultivation: Proceedings of the Second International Conference on Edible Mycorrhizal Mushrooms*. Christchurch, New Zealand, 3–5 July 2001. CD-ROM. Christchurch, New Zealand Institute for Crop and Food Research Limited.

Honrubia, M., Morte, A., Gutiérrez, A. 2005. Six year of the *Terfezia claveryi* cultivation in Murcia (Spain). In: *Abstracts of the Fourth International Workshop on Edible Mycorrhizal Mushrooms*. Murcia, Spain, 28 November–2 December 2005. Universidad de Murcia, Murcia. P. 70.

Honrubia, M., Torres, P., Diaz, G., Cano, A. 1992. *Manual para micorrizar plantas en viveros forestales.* Ministerio de Agricultura. Instituto para la Conservación de la Naturaleza (ICONA), Madrid.

Honrubia, M., Torres, P., Morte, A. 1993. *Biotecnología forestal: micorrización y micro-propagación.* Universidad de Murcia, Centro Internacional de Altos Estudios Agronómicos Mediterráneos.

Hooke, R. 1665. *Micrographia.* Royal Society, London. Available via www.gutenberg. org/files/15491/15491-h/15491-h.htm.

Horton, T.R., Bruns, T.D. 2001. The molecular revolution in ectomycorrhizal ecology: peeking into the black box. *Molecular Ecology* 10:1855–1871.

How Much Is That Worth Today? 2005. eh.net/hmit/ppowerbp/.

Hu, H.-T. 1992. *Tuber formosanum* sp. nov. and its mycorrhizal associations. *Journal of Experimental Forestry* (National Taiwan University) 6:79–86.

Hume, D. 1996. *Bechamp or Pasteur: A Lost Chapter in the History of Biology.* Kessinger, Whitefish, Mont.

Iddison, P. 2000. Desert truffles *Tirmania nivea* in the Emirates. *Tribulus Magazine* 10(1):20–21. Available via www.enhg.org/trib/trib10.htm.

Iddison, P. 2004. Truffles in Middle Eastern cookery. Available via enhg.4t.com/iddison/destruf.htm

Imazeki, R., Otani, Y., Hongo, T., Izawa, M., Mizuno, N. 1988. *Coloured Illustrations of Mushrooms of Japan.* Yama-kei, Tokyo (in Japanese).

Innvista. 2006. Truffles. Available via www.innvista.com/health/foods/mushrooms/truffle.htm.

Institut Recherché National Agronomique. 2005. La truffe: de plus en plus rare et chère, gare aux fraudes! *Fiche de Presse Info*, 14 December. Available via www.inra. fr/presse/la_truffe_de_plus_en_plus_rare_et_chere_gare_aux_fraudes.

Integrated Pest Management of Alaska. 2003. Powdery mildew of roses. Available via www.ipmofalaska.com/files/powderymildewroses.html.

International Culture Collection of Arbuscular and Vesicular-Arbuscular Mycorrhizal Fungi. 2006. Classification of Glomeromycota. Available via www.invam. caf.wvu.edu/fungi/taxonomy/classification.htm.

Iotti, M., Amicucci, A., Stocchi, V., Zambonelli, A. 2002. Morphological and molecular characterization of mycelia of some *Tuber* species in pure culture. *New Phytologist* 155:499–505.

Iotti, M., Zambonelli, A. 2005. A quick and precise technique for identifying ectomycorrhizas by PCR. *Mycological Research* 110:60–65.

Iqbal, M. 1993. *International Trade in Non-wood Forest Products.* FAO Miscellaneous Publication no. 93/11. Food and Agriculture Organization of the United Nations, Rome.

Istituto Nazionale di Statistica. 1990–2006. *Bollettino mensile di statistica.* Available via www.istat.it.

Istrianet.org. 2004. Truffles-tartufi: the white truffle in Istria. Available via www. istrianet.org/istria/flora/fungi/truffles-istria.htm.

Jacobson, P. 1990. Sleuths sniff out truffle faking ring. *Times* (London), 6 January.

James, B. 2003. French use genetic coding tools to root out truffle fraud. *International Herald Tribune*, 11 January. Available via www.iht.com/articles/2003/01/11/truffle_ed3_php.

Jarvis, P., Warren, I., Hall, I.R. 1988. Investment in truffière development. In: *New Zealand Ministry of Agriculture and Fisheries South Region Business Plan*. New Zealand Ministry of Agriculture and Fisheries, Invermay.

Johansson, J. 2001. Ecology and control of oak mildew (*Microsphaera alphitoides*). Available via www.mykopat.slu.se/Newwebsite/mycorrhiza/kantarellfiler/texter/mildew.html.

Johnston, B. 2000. Gold is not worth its weight in truffles. *Daily Telegraph*, 14 November.

Joint FAO/WHO Expert Committee on Food Additives. 2001. Summary of evaluations performed by the Joint FAO/WHO Expert Committee on Food Additives: bis(methylthio)methane. Available via www.jecfa.ilsi.org/evaluation.cfm?chemical=bis(METHYLTHIO)METHANE&keyword=FLAVOURING.

Jones, J.B., Wolf, B., Mills, H.A. 1991. *Plant Analysis Handbook: A Practical Sampling, Preparation, Analysis, and Interpretation Guide*. Micro-Macro Publishing, Athens, Ga.

Jongbloed, M. 2005. Desert truffles, a disappearing delicacy. *Al Shindagah* March–April. Available via www.alshindagah.com/marapr2005/dessert.html.

Kagan-Zur, V., Kuang, J., Tabak, S., Taylor, F.W., Roth-Bejerano, N. 1999. Potential verification of a host plant for the desert truffle *Terfezia pfeilii* by molecular methods. *Mycological Research* 103:1270–1274.

Kagan-Zur, V., Wenkart, S., Mills, D., Freeman, S., Luzzati, Y., Ventura, Y., Zaretsky, M., Roth-Bejerano, N., Shabi, E. 2002. *Tuber melanosporum* research in Israel. In: Hall, I.R., Wang, Y., Danell, E., Zambonelli, A., eds. *Edible Mycorrhizal Mushrooms and Their Cultivation: Proceedings of the Second International Conference on Edible Mycorrhizal Mushrooms*. Christchurch, New Zealand, 3–5 July 2001. CD-ROM. New Zealand Institute for Crop and Food Research Limited, Christchurch.

Kantonalen Laboratorium. 2005. Olive oils with truffle flavouring: declaration and flavourings. Available via www.kantonslabor-bs.ch/files/berichte/Report0455.pdf.

KBC Tree Shelters. 2006. www.killyleaghbox.co.uk/horticulture.htm.

Keating, G. 2002. Gourmet pays $35,000 for truffle. *Excite News*. Available via www.news.excite.com/odd/article/id/280829%7Coddlyenough%7C11-12-2002::08:51%7Creuters.html.

Kendrick, B. 2002. *The Fifth Kingdom on CD-ROM*. Mycologue Publications, Sidney. Available via www.mycolog.com/fifthtoc.html.

Kers, L.E. 2003. Tryfflarna *Tuber aestivum* och *T. mesentericum* i Sverige (*Tuber aestivum* Vitt. and *T. mesentericum* Vitt. in Sweden). Svensk Botanisk Tidskrift 97:157–175.

Khabar, L., Najim, L., Janex-Favre, M.-C. 2001. Contribution a l'étude de la flore mycologique du Maroc les Marocaines (Discomycètes). *Bulletin société mycologique France* 117:213–229.

Khabar, L., Slama, A., Neffati, M. 2005. Terfess common to Morocco and Tunisia. In: *Proceedings of the Fourth International Workshop on Edible Ectomycorrhizal Mushrooms*. Murcia, Spain, 28 November–2 December 2005. P. 75.

Khare, K.B. 1975. *Terfezia terfezioides*: a new record for India. *Current Science* 44:601–602.

Kirk, P.M., David, J.C., Staplers, J.A., eds. 2001. *Ainsworth and Bisby's Dictionary of the Fungi*. CAB International, Wallingford.

Kovács, G.M. 2002. Study of mycorrhizae on the Great Hungarian Plain. Ph.D. diss., University of Szeged, Hungary. Available via www.vmri.hu/~gkovacs/engthes.doc.

Kovács, G.M., Vágvölgy, C., Oberwinkler, F. 2003. *In vitro* interaction of the truffle *Terfezia terfezioides* with *Robinia pseudoacacia* and *Helianthemum ovatum*. *Folia microbiologica* 48:369–378.

Kropp, B.R., Mueller, G.M. 2000. *Laccaria*. In: Cairney, J.W.G., Chambers, S.M., eds. *Ectomycorrhizal Fungi: Key Genera in Profile*. Springer Verlag, Berlin. Pp. 65–88.

Kuepper, G., Thomas, R., Earles, R. 2001. Use of baking soda as a fungicide. National Sustainable Agriculture Information Service. Available via www.attra.ncat.org/attra-pub/PDF/bakingsoda.pdf.

Kues, U. 2000. Life history and developmental processes in the basidiomycete *Coprinus cinereus*. *Microbiology and Molecular Biology Reviews* 64:315–353.

Kunming Yunri Foods. 2006. Chinese truffle. Available via www.sinohost.com/yunnan_pages/mushrooms/chinesetruffle.html.

Kuwait Information Office. 2005. Impact of Gulf War. Available via www.kuwait-info.com/sidepages/environment_impactofgulf.asp.

Laatikainen, T., Heinonen-Tanski, H. 2002. Mycorrhizal growth in pure cultures in the presence of pesticides. *Microbiological Research* 157:127–137.

Lacroix, P. 2004. Manners, custom and dress during the Middle Ages and during the Renaissance period. Project Gutenberg ebook. Available via www.gutenberg.org/etext/10940.

Lambert, J.M., Manners, J.M., Westrup, A.W., Paton, J.A., Hora, F.B., Blaikley, N.M., Bradshaw-Bond, M. 1964. Botany. In: F.J. Monkhouse, ed. *A Survey of Southampton and Its Region*. Southampton University Press, Southampton. Available via www.hants.gov.uk/newforesthistory/botany.htm.

Lanfranco, L., Wyss, P., Marzachi, C., Bonfante, P. 1993. DNA probes for identification of the ectomycorrhizal fungus *Tuber magnatum* Pico. *FEMS Microbiology Letters* 114(3):245–251.

Lange, C. 2001. Status over Sommer-Trøffel i Danmark. *Svampe* 43:5–8.

Lännen Plant Systems. 2006. www.lannenplantsystems.com/?id=3CC2724D-72B242BB0A02-B622DEA925B4.

Lansac, A.R., Marín, A., Roldán, A. 1995. Mycorrhizal colonization and drought interactions of Mediterranean shrubs under greenhouse conditions. *Arid Soil Research and Rehabilitation* 9:167–175.

Lawrynowicz, M. 1993. Distributional limits of truffles in northern Europe. *Micologia e Vegetazione Mediterranea* 7(1):31–38.

Leake, J.R., Johnson, D., Donnelly, D., Muckle, G.E., Boddy, L., Read, D.J. 2004. Networks of power and influence: the role of mycorrhizal mycelium in controlling plant communities and agro-ecosystem functioning. *Canadian Journal of Botany* 82:1015–1045.

Lee, R.B. 1894. *A History and Description of the Modern Dogs of Great Britain and Ireland*. Horace Cox, London. Pp. 185–187.

Lefevre, C., Hall, I.R. 2001. The global status of truffle cultivation. In: Mehlenbacher, S.A., ed. Fifth International Congress on Hazelnut, Corvallis, Oregon, August 2000, International Society for Horticultural Science. *Acta Horticulturae* 556:513–520.

Leffers, A. 2003. *Gemsbok Bean and Kalahari Truffle: Traditional Plant Use by Jul'hoansi in North-eastern Namibia*. Gamwberg Macmillan, Windhoek.

Legge quadro nazionale no. 752. 1985. (Quality standards for preserved truffles.) Available via www.digilander.libero.it/dlfrimini/micologia/leggi/italia/legge_752_1985.htm.

Legge quadro nazionale no. 162. 1991. (Quality standards for preserved truffles.) Available via www.digilander.libero.it/dlfrimini/micologia/leggi/italia/legge_162_1991.htm#art.1.

Lendering, J. 2005. Babylonia: country, language, religion, culture. Available via www.livius.org/ba-bd/babylon/babylonia.html.

Lepp, H., Fagg, M. 2004. Mycogeography: imports, exports and puzzles. Available via the Australian Botanic Gardens fungi web site, www.anbg.gov.au/fungi/mycogeography-imports.html.

Le Trufficulteur. (A French magazine on truffle growing published four times a year.) www.fft-tuber.org.

Liguria legge regionale. 2006. leggi.regione.liguria.it/leggi/leggiric.htm.

Lincoff, G.H., Nehring, C. 1995. *The Audubon Society Field Guide to North American Mushrooms*. Knopf, New York.

Linés Escardó, A. 1970. The climate of the Iberian Peninsula. In: Wallén, C.C., ed. *World Survey of Climatology*. Vol. 5, *Climates of Northern and Western Europe*. Elsevier, Amsterdam. Pp. 195–239.

Linnaeus, C. 1753. *Species Plantarum*. 2 vols. Reprint 1957–1959. Ray Society, London.

Little, R.C. 1971. The treatment of iron deficiency. In: *Trace Elements in Soil and Crops*. Ministry of Agriculture and Fisheries Technical Bulletin no 21. Her Majesty's Stationery Office, London. Pp. 45–61.

Lo Bue, G., Gregori, G.L., Pasquini, L., Maggiorotto, G. 1990. Sintesi micorrizica in campo fra piante adulte e tartufi pregiati mediante frammenti radicali. In: Bencivenga, M., Granetti, B., eds. *Atti del secondo congresso internazionale sul tartufo*. Spoleto, Italy, 24–27 November 1988. Comunità Montana dei Monti Martani e del Serano, Spoleto. Pp. 459–466.

Longato, S., Bonfante, P. 1997. Molecular identification of mycorrhizal fungi by direct amplification of microsatellite regions. *Mycological Research* 101:425–432.

Louisiana University AgCenter. 2005. Insect and disease control: Appendix—trade names of fungicides and nematodes listed alphabetically. Available via www.lsuagcenter.com/NR/rdonlyres/FC1EC609-B31A-48CB-8438-475DB21DC8E9/24146/AppendixTradeNames.pdf.

Love to Know. 1911. Silk. Online encyclopaedia, accessed 2007. Available via www.encyclopedia.org/S/SI/SILK.htm.

Lulli, L., Bragato, G., Gardin, L. 1999. Occurrence of *Tuber melanosporum* in relation to soil surface layer properties and soil differentiation. *Plant and Soil* 214:85–92.

Lulli, L., Bragato, G., Gardin, L., Panini, T., Primavera, F. 1992. I suoli delle tartufaie naturali della bassa valle del Santerno (Mugello-Toscana). *L' Italia forestale e montana* 5:251–267.

Lulli, L., Pagliai, M., Bragato, G., Primavera, F. 1993. La combinazione dei caratteri che determinano il pedoambiente favorevole alla crescita del *Tuber magnatum* Pico nei suoli dei depositi marnosi dello Schlier in Acqualagna (Marche). *CNR Quaderni di Scienza del Suolo* 5:143–159.

Lulli, L., Panini, T., Bragato, G., Gardin, L., Primavera, F. 1991. I suoli delle tartufaie naturali delle Crete Senesi. *Monti e Boschi* 42(5):17–24.

Lulli, L., Primavera, F. 1995. I suoli idonei alla produzione di tartufi. *L'informatore agrario* 51:33–38.

Luoma, D.L., Eberhart, J.L., Molina, R., Amaranthus, M.P. 2004. Response of ectomycorrhizal fungus sporocarp production to varying levels and patterns of greentree retention. *Forest Ecology and Management* 202:337–354.

Luppi-Mosca, A.M., Fontana, A. 1977. Researches on *Tuber melanosporum* ecology. IV. Mycological analyses of central Italy truffle soils. *Allionia* 22:105–114.

Mabru, D., Dupre, C., Douet, J.P., Leroy, P., Ravel, C., Ricard, J.M., Medina, B., Castroviejo, M., Chevalier, G. 2001. Rapid molecular typing method for the reliable detection of Asiatic black truffle (*Tuber indicum*) in commercialized products: fruiting bodies and mycorrhizal seedlings. *Mycorrhiza* 11(2):89–94.

MacFarquhar, N. 2004. Beneath desert sands, an Eden of truffles. Available via www.natruffling.org/desert.htm.

Maclaren, J.P. 1993. *Radiata Pine Grower's Manual*. Forest Research Institute Bulletin no. 184. New Zealand Forest Research Institute, Rotorua.

Malajczuk, N., Molina, R., Trappe, J. 1982. Ectomycorrhiza formation in *Eucalyptus*. I. Pure culture synthesis, host specificity and mycorrhizal compatibility with *Pinus radiata*. *New Phytologist* 91:467–482.

Malajczuk, N., Reddell, P., Brundrett, M. 1994. Role of ectomycorrhizal fungi in minesite reclamation. In: Pfleger, F.L., Linderman, R.G., eds. Mycorrhizae and Plant Health. American Phytopathological Society, St. Paul, Minn. Pp. 83–100.

Malençon, M.G. 1938. Les truffes Européennes: historique, morphogenie, organographe, classification, culture. *Revue de Mycologie: Mémoire hors-série* 1:1–92.

Mamoun, M., Olivier, J.M. 1990. Dynamique des populations fongiques et bactériennes de la rhizosphére des noisetiers truffiers. III. Effet du régime hydrique sur la mycorhization et la microflore associée. *Agronomie* 10:77–84.

Mamoun, M., Olivier, J.M. 1992. Effect of soil pseudomonads on colonization of hazel roots by the ecto-mycorrhizal species *Tuber melanosporum* and its competitors. *Plant and Soil* 139:265–273.

Mamoun, M., Olivier, J.M. 1996. Receptivity of cloned hazels to artificial ectomycorrhizal infection by *Tuber melanosporum* and symbiotic competitors. *Mycorrhiza* 6:15–19.

Mamoun, M., Olivier, J.M. 1997. Mycorrhizal inoculation of cloned hazels by *Tuber*

melanosporum: effect of soil disinfestation and co-culture with *Festuca ovina*. *Plant and Soil* 188:221–226.

Mannozzi-Torini, L. 1976. *Manuale di tartuficoltura: tartufi e tartuficoltura in Italia*. Edagricole, Bologna.

Mannozi-Torini, L. 1984. *Il Tartufo e la sua coltivazione*. Edagricole, Bologna.

Marchand, A. 1971–1986. *Champignons du nord et du midi*. Vols. 1–9. Société Mycologique des Pyrénées Méditerranéennes, Perpignan, France.

Marino, R., Cerone, G., Rana, G.L. 2003. Studi sui funghi ipogei della Basilicata. II. *Rivista di micologia* 46(1):53–62.

Maser, C., Trappe, J.M., Nussbaum, R.A. 1978. Fungal–small mammal interrelationships with emphasis on Oregon coniferous forests. *Ecology* 59:799–809.

Mason, P.A., Wilson, J., Last, F.T., Walker, C. 1983. The concept of succession in relation to the spread of sheathing mycorrhizal fungi on inoculated tree seedlings growing in unsterile soil. *Plant and Soil* 71:247–256.

Matruchot, L. 1903. *Germination des spores des truffes, colture et caractères du mycelium truffier*. Comptes rendus de l'Academie des sciences, Paris.

Mauriello, G., Marino, R., D'Auria, M., Cerone, G., Rana, G.L. 2004. Determination of volatile organic compounds from truffles via SPME-GC-MS. *Journal of Chromatographic Science* 42:299–305.

Maybury, G. 2002. *Pig Apples*. Scholastic, Auckland.

Mayle, P. 1990. *A Year in Provence*. Pan, London.

Mayle, P. 1992. *Toujours Provence*. Pan, London.

Mayle, P. 1999. *Encore Provence*. Hamilton, London.

Mayuzumi, Y., Mizuno, T. 1997. Cultivation methods of maitake (*Grifola frondosa*). *Food Reviews International* 13:357–364.

McRae, J. 1999. Commercial containerized hardwood seedling production in the southern USA. In: Landis, T.D., Barnett, J.P., eds. *National Proceedings: Forest and Conservation Nursery Associations, 1998*. General Technical Report no. SRS-25. U.S. Department of Agriculture Forest Service, Southern Research Station, Asheville, N.C. Pp. 35–38.

Mead, G. 2006. Hazel coppice. Available via www.hazelwattle.com/hazelcoppice.html.

Mello, A., Cantisani, A., Vizzini, A., Bonfante, P. 2002. Genetic variability of *Tuber uncinatum* and its relatedness to other black truffles. *Environmental Microbiology* 4:584–594.

Mello, A., Fontana, A., Meotto, F., Comandini, O., Bonfante, P. 2001. Molecular and morphological characterization of *Tuber magnatum* mycorrhizas in a long-term survey. *Microbiological Research* 155:279–284.

Mello, A., Garnero, L., Bonfante, P. 1999. Specific PCR primers as a reliable tool for the detection of white truffles in mycorrhizal roots. *New Phytologist* 141(3):511–516.

Mello, A., Murat, C., Vizzini, A., Gavazza, V., Bonfante, P. 2005. *Tuber magnatum* Pico, a species of limited geographical distribution: its genetic diversity inside and outside a truffle ground. *Environmental Microbiology* 7:55–65.

Mello, A., Vizzini, A., Longato, S., Rollo, F., Bonfante, P., Trappe, J.M. 2000. *Tuber borchii* versus *T. maculatum*: neotype studies and DNA analyses. *Mycologia* 92:326–333.

Meotto, F., Carraturo, P. 1987–1988. Ectomicorrizia di *Sphaerospora brunnea* (A. and S.) Svrcek and Kubicka in piantine tartufigene. *Allionia* 28:109–116.

Meotto, F., Carraturo, P., Dana, M. 1992. Valutazione in pieno campo e in serra della competitività di *Sphaerosporella brunnea* con *Tuber magnatum*. *L'informatore agrario* 48(47):73–78.

Met Office. 2007. www.met-office.gov.uk/climate/uk/averages/19712000/index.html.

Micheli, P.A. 1729. *Nova plantarum genera juxta Tournafortii methodum disposita.* Florence.

Mirabella, A., Primavera, F., Gardin, L. 1992. Formation dynamics and characterization of clay minerals in a natural truffle bed of *Tuber magnatum* Pico on Pliocene sediments in Tuscany. *Agricoltura Mediterranea* 122:275–281.

Mitchell, J., Zuccaro, A. 2006. Sequences, the environment, and fungi. *Mycologist* 20:62–74.

Mitrovic, M., Milenkovic, M., Pavlovic, P., Djurdjevic, L. 2006. Osmotic potential and water content in fruit bodies of white truffle (*Tuber magnatum* Pico) in two different forest soil in Serbia. In: Khabar, L., ed. *Le premier symposium sur les champignons hypogés du basin Méditerranéen.* Rabat, Morocco, 5–8 April 2004. Pp. 136–145.

Miyauchi, S., Kon, K., Yamauchi, T., Shimomura, M. 1998. Cultural characteristics of mycelial growth of *Pleurotus eryngii*. *Nippon-kingakukai-kaiho* 39:83–87.

Molina, R., O'Dell, T., Luoma, D., Amaranthus, M., Castellano, M., Russel, K. 1993. Biology, ecology, and social aspects of wild edible mushrooms in the forests of the Pacific Northwest: a preface to managing commercial harvest. Forest Service, Pacific Northwest General Technical Report no. 309. U.S. Department of Agriculture Forest Service, Portland, Ore.

Molina, R., Palmer, J.G. 1982. Isolation, maintenance, and pure culture manipulation of ectomycorrhizal fungi. In: Schenck, N.C., ed. *Methods and Principles of Mycorrhizal Research.* American Phytopathological Society, St. Paul, Minn. Pp. 115–129.

Montagne, P. 1961. *Larousse gastronomique.* Froud, N., Turgeon, C., eds. Hamlyn, London.

Montecchi, A., Sarasini, M. 2000. *Funghi ipogei d'europa.* Associazione Micologica Bresadola, Trento.

Moore, D. 1998. *Fungal Morphogenesis.* Cambridge University Press, Cambridge.

Moore, I. 1985. *The Truffle Hunter.* Arrow Books (Beaver Books), London.

Morcillo, M., Sánchez, M., Garcia, E. 2005. Open field inoculation of adult hazel groves with *Tuber melanosporum* Vitt. In: *Proceedings of the Fourth International Workshop on Edible Mycorrhizal Mushrooms.* Murcia, Spain, 28 November–2 December 2005. Universidad de Murcia, Murcia.

Moreno, G., Díez, J., Manjón, J.L. 2000. *Picoa lefebvrei* and *Tirmania nivea*, two rare hypogeous fungi from Spain. *Mycological Research* 104:378–381.

Moreno, G., Díez, J., Manjón, J.L. 2002. *Terfezia boudieri*, first records from Europe of a rare vernal hypogeous mycorrhizal fungus. *Persoonia* 17:637–641.

Moreno, G., Manjón, J.L., Díez, J. 1997. *Tuber pseudohimalayense* sp. nov.: an Asiatic species commercialized in Spain, similar to the Périgord black truffle. *Mycotaxon* 68:217–224.

Moreno Arroyo, B., Gómez Fernández, J., Pulido Calmaestra, E. 2005. *Tesores de nuestros montes: Trufas de Andalucía*. Consejería de medio ambiente, Junta de Andalucía.

Moriondo, F., Capretti, P., Ragazzi, A. 2006. *Malattie delle piante in bosco, in vivaio e delle alberature*. Quarto inferiore, Bologna.

Morte, M.A., Honrubia, M. 1994. Método para la micorrización in vitro de plantas micropropagadas de *Helianthemum* con *Terfezia claveryi*. Spanish patent P9402430.

Mshigeni, K.E. 2001. The cost of scientific and technological ignorance with special reference to Africa's rich biodiversity. UNDP/UNOPS Regional Project RAF/99/021. ZERI Regional Office for Africa. Available via www.zeri.unam.na/Text/THeCostOFScientific.htm.

Murat, C., Díez, J., Luis, P., Delaruelle, C., Dupré, C., Chevalier, G., Bonfante, P., Martin, F. 2004. Polymorphism at the ribosomal DNA ITS and its relation to postglacial re-colonization routes of the Périgord truffle *Tuber melanosporum*. *New Phytologist* 164:401–411.

Murat, C., Vizzini, A., Bonfante, P., Mello, A. 2005. Morphological and molecular typing of the below-ground fungal community in a natural *Tuber magnatum* truffle-ground. *FEMS Microbiology Letters* 245:307–313.

Mushroom Company. 2007. U.S. wholesale market prices. *Mushroom Growers' Newsletter* (Klamath Falls, OR).

Nagy, M. 1988. Untersuchung von getrüffelten Fleischerzeugnissen auf Trüffeln. *Fleischwirtschaft* 68:592–593.

Namibia. 2002. Initial national communication to the United Nations framework convention on climate change. Available via www.unfccc.int/resource/docs/natc/namnc1.pdf.

Nanagulian, S.G., Senn-Irlet, B. 2002. Some dates about distribution and conservation of threatened mushrooms in Armenia. Available via www.wsl.ch/eccf/Armenia.pdf.

Natale, D., Pasqualini, E. 1999. Control of *Zeuzera pyrina* and *Cossus cossus* using pheromones. *L'informatore agrario* 55:79–83.

Natarajan, K., Mohan, V., Ingleby, K. 1992. Correlation between basidiomata production and ectomycorrhizal formation in *Pinus patula* plantations. *Soil Biology and Biochemistry* 24:279–280.

National Institute of Water and Atmospheric Research. 2007. www.niwa.co.nz/ncc.

NationalPak. 2006. The role of micronutrients in crop production. Available via www.nationalpak.com/roleofmicro.asp.

Newton, A.C., Haigh, J.M. 1998. Diversity of ectomycorrhizal fungi in Britain: a test of the species-area relationship, and the role of host specificity. *New Phytologist* 138:619–627.

New Zealand Department of Scientific and Industrial Research, Soils Bureau. 1968. *General Survey of the Soils of South Island, New Zealand: Wellington*. New Zealand Department of Scientific and Industrial Research Soil Bureau Bulletin no. 27.

New Zealand Meteorological Service. 1980. *Summaries of Climatological Observations*

to 1980. New Zealand Meteorological Service Miscellaneous Publication no. 177, Wellington.

New Zealand Ministry of Agriculture and Forestry. 2004. Importation into New Zealand of specified fresh and frozen *Tuber* species (truffles). MAF Biosecurity Authority (Plants) Standard PIT-IMP-TUBER. Available via www.biosecurity. govt.nz/imports/plants/standards/pit-imp-tuber.pdf.

New Zealand Plant Conservation Network. 2003. *Pennantia baylisiana* (W.R.B. Oliv.) G.T.S. Baylis. Available via www.nzpcn.org.nz/nz_threatenedplants/ advanced_search.asp.

North American Truffling Society. 1987. *The Cookbook of North American Truffles.* North American Truffling Society, Corvallis, Ore.

O'Donnel, K., Cigelnik, E., Weber, N.S., Trappe, J.M. 1997. Phylogenetic relationships among ascomycetous truffles and the true and false morels inferred from 18S and 28S ribosomal DNA sequence analysis. *Mycologia* 89:48–65.

Ohenoja, E., Lahti, S. 1978. Food from the Finnish forests. *Proceedings of the Eighth World Forestry Congress.* Vol. 3, *Forestry for Food.* Pp. 1013–1021.

Olivier, J.M. 2000. Progress in the cultivation of truffles. In: Van Griensven, L.J.L.D. ed. *Mushroom Science XV: Science and Cultivation of Edible Fungi.* Vol. 2. Balkema, Rotterdam. Pp. 937–942.

Olivier, J.M., Delmas, J. 1987. Vers la maitrise des champignons comestibles. *Biofutur* Octobre:23–41.

Olivier, J.M., Savignac, J.C., Sourzat, P. 2002. *Truffe et trufficulture.* Fanlac, Périgueux, France.

Olney, R. 1985. *The French Menu Cookbook.* Godine, Boston.

Olsen, J. 2001. Nutrient management guide: hazelnuts. Oregon State University Extension Service. Available via http://eesc.orst. edu/agcomwebfile/edmat/html/EM/EM8785-E/EM8785-E.html.

Olsson, P.A., Münzenberger, B., Mahmood, S., Erland, S. 2000. Molecular and anatomical evidence for a three-way association between *Pinus sylvestris* and the ectomycorrhizal fungi *Suillus bovinus* and *Gomphidius roseus. Mycological Research* 104:1372–1378.

Oregon White Truffles. 2006. A history of Oregon white truffles. Available via www. oregonwhitetruffles.com/.

Owen, R. 2000. £5,000 paid for one truffle in Italy's white gold rush. *Times* (London), 14 November, p. 14.

Pacioni, G. 1985. *La coltivazione moderna e redditizia del tartufo: Guida practica.* Giovanni De Vecchi Editore S.p.A., Milano.

Pacioni, G. 1989. Biology and ecology of the truffles. *Acta Medica Romana* 27:104–117.

Pacioni, G., Bologna, M.A., Laurenzi, M. 1991. Insect attraction by *Tuber*: a chemical explanation. *Mycological Research* 95:1359–1363.

Pacioni, G., Comandini, O. 2000. *Tuber.* In: Cairney, J.W.G., Chambers, S.M., eds. *Ectomycorrhizal Fungi: Key Genera in Profile.* Springer Verlag, Berlin. Pp. 163–186.

Pacioni, G., Marra, L., eds. 1993. *Tuber.* Atti del convegno internazionale sul tartufo. *Micologia e Vegetazione Mediterranea 7.*

Palenzona, M. 1969. Sintesi micorrizica tra *Tuber aestivum* Vitt., *Tuber brumale* Vitt., *Tuber melanosporum* Vitt. e semenzali di *Corylus avellana* L. *Allionia* 15:121–131.

Palenzona, M., Chevalier, G., Fontana, A. 1972. Sintesi micorrizica tra i miceli in coltura di *Tuber brumale, T. melanosporum, T. rufum* e semenzali di conifere e latifoglie. *Allionia* 18:41–52.

Palenzona, M., Curto, A., Mondino, G.P., Saladin, R. 1976. *Il tartufo di Bagnoli Tuber mesentericum Vitt.* Camera di Commercio Industria, Artigianato e Agricoltura, Avellino.

Panchuk, K., Karamanos, R., Mahli, S.S., Flaten, P. 2000. Micronutrients in crop production. Available via www.agr.gov.sk.ca/DOCS/production/micronutrients. asp.

Paolocci, F., Angelini, P., Cristofari, E., Granetti, B., Arcioni, S. 1995. Identification of *Tuber* spp. and corresponding ectomycorrhizae through molecular markers. *Journal of Science and Food Agriculture* 69:511–517.

Paolocci, F., Rubini, A., Granetti, B., Arcioni, S. 1997. Typing *Tuber melanosporum* and Chinese black truffle species by molecular markers. *FEMS Microbiology Letters* 153(2):255–260.

Paolocci, F., Rubini, A., Granetti, B., Arcioni, S. 1999. Rapid molecular approach for a reliable identification of *Tuber* spp. ectomycorrhizae. *FEMS Microbiology Ecology* 28:1, 23–30.

Paolocci, F., Rubini, A., Riccioni, C., Arcioni, S. 2006. Re-evaluation of the life cycle of *Tuber magnatum. Applied and Environmental Microbiology* 72:2390–2393.

Paolocci, F., Rubini, A., Riccioni, C., Topini, F., Arcioni, S. 2004. *Tuber aestivum* and *Tuber uncinatum*: Two morphotypes or two species? *FEMS Microbiology Letters* 235:109–115.

Papa, G. 1980. Purification attempts of the plant inhibitory principle of *Tuber melanosporum* Vitt. *Phytopatologia Mediterranea* 19:177.

Parladé, J., Pera, J., Alvarez, I.F. 1996. Inoculation of containerized *Pseudotsuga menziesii* and *Pinus pinaster* seedlings with spores of five species of ectomycorrhizal fungi. *Mycorrhiza* 6:237–245.

Payne, S. 2004. Berkshire truffles valued at £33,000. *Daily Telegraph*, 24 September. Available via www.telegraph.co.uk/news/main.jhtml?xml=/news/2004/09/24/ntruff24.xml.

Pébeyre, P.-J., Gleyze, R., Montant, C. 1985. Product for the fertilization of mycorrhizal mushrooms and application to the fertilization of truffle-beds. U.S. patent 4,537,613.

Pébeyre, P.-J., Pébeyre, J. 1987. *Le grand livre de la truffe.* Briand-Laffont, Paris.

Pecan Truffles. 2003. Pecan truffle fact sheet. Available via www.interests.caes.uga. edu/pecantruffles/factsheet.htm.

Peer, E. 1980. On the trail of the truffle. *Geo* 2(5):112–130.

Pegler, D.N. 2002. Useful fungi of the world: the 'poor man's truffles of Arabia' and 'manna of the Israelites'. *Mycologist* 16:8–9.

Pegler, D.N., Spooner, B.M., Young, T.W.K. 1993. *British Truffles: A Revision of British Hypogeous Fungi.* Royal Botanic Gardens, Kew.

Pelusio, F., Nilsson, T., Montanarella, L., Tilio, R., Larsen, B., Facchetti, S., Mad-

sen, J.-O. 1995. Headspace solid-phase microextraction analysis of volatile organic sulfur compounds in black and white truffle aroma. *Journal of Agricultural and Food Chemistry* 43:2138–2143.

Percurdani, R., Trevisi, A., Zambonelli, A., Ottonello, S. 1999. Molecular phylogeny of truffles (Pezizales: Terfeziaceae, Tuberaceae) derived from nuclear rDNA sequence analysis. *Molecular Phylogenetics and Evolution* 13:169–180.

Perrott, K. 2004. How do soil P tests relate to soil phosphorus? Hill Laboratories Science Views 3. Available via www.hill-labs.co.nz/Files/PDFs/ScienceViewsIssue3.pdf.

Peterson, R.L., Massicotte, H.B., Melville, L.H. 2004. *Mycorrhizas: Anatomy and Cell Biology.* NRC Press, Ottawa.

Petit, R.J., Brewer, S., Bordacs, S., Burg, K., Cheddadhi, R., Coan, E., Cottrell, J., Qaild, U.M., Van Dam, B., Deans, J.D., Espinel, S., Finschi, S., Finkeldey, R., Glaz, I., Giocoechea, P.G., Jensen, J.S., Konig, A.O., Lowe, A.J., Madsen, S.F., Matyas, G., Munro, M.C., Popescu, F., Slade, D., Tabbener, H., deVries, S.G.M., Ziegenhagen, B., de Beaulieu, J.L., Kremer, A. 2002. Identification of refugia and post-glacial colonisation routes of European white oaks based on chloroplast DNA and fossil pollen evidence. *Forest Ecology and Management* 156:49–74.

Petrarca, F. 2000. *Canzoniere.* Edited and translated by J.G. Nichols. Carcanet Press, Manchester.

Petrarch, F. 2006. *The Canzoniere.* Edited and translated by A.S. Kline. Available via www.petrarch.petersadlon.com/canzoniere.html.

Pettenella, D., Klöhn, S., Brun, F., Carbone, F., Venzi, L., Cesaro, L., Ciccarese, L. 2004. Economic integration of urban consumers' demand and rural forestry production. Italy's Country Report, COST Action E30. Available via www.apat.gov.it/site/_files/English_documents/Italy-Report.pdf.

Peverill, K.I., Sparrow, L.A., Reuter, D.J. 2001. *Soil Analysis: An Interpretation Manual.* CSIRO, Collingwood, Austrailia.

PheroTech International. 2006. Pheromone technology. Available via www.pherotech.xplorex.com/page193.htm.

Phillips, R. 1981. *Mushrooms and Other Fungi of Great Britain and Europe.* Pan, London.

Picart, F. 1980. *Truffle: The Black Diamond.* Agri-Truffle, Santa Rosa, Calif.

Pilz, D., Smith, J., Amaranthus, M.P., Alexander, S., Molina, R., Luoma, D. 1999. Mushrooms and timber: managing commercial harvesting in the Oregon Cascades. *Journal of Forestry* 97(3):4–11.

Pinkas, Y., Maimon, M., Shabi, E., Elisha, S., Shmulewich, Y., Freeman, S. 2000. Inoculation, isolation and identification of *Tuber melanosporum* from old and new oak hosts in Israel. *Mycological Research* 104:472–477.

Pirazzi, R. 1990. Produzione naturale di *Tuber* spp. in rimboschimenti di cedro e prove di sintesi. In: Bencivenga, M., Granetti, B., eds. *Atti del secondo congresso internazionale sul tartufo.* Spoleto, Italy, 24–27 November 1988. Comunità Montana dei Monti Martani e del Serano, Spoleto. Pp. 303–311.

Pirazzi, R., Gregorio, A. 1987. Accrescimento di conifere micorrizate con specie diverse di *Tuber. Micologia Italiana* 16(3):49–62.

Plant Health Risk Assessment Unit. 2003. Hosts of *Phytophthora ramorum* with notes on geographical distribution and mating types, causal agent of sudden oak death. Available via www.nature.berkeley.edu/comtf/pdf/P.ramorum.hosts. June.2003.pdf.

Plattner, I., Hall, I.R. 1995. Parasitism of non-host plants by the mycorrhizal fungus *Tuber melanosporum* Vitt. *Mycological Research* 99:1367–1370.

Poitou, N. 1986. *Le sol: Cas particulier des sols truffiers.* INRA, Pont-de-la-Maye Station de Recherches sur les Champignons, Stage CTIFL, 1985–1986.

Poitou, N. 1990. Les sols truffiers français. In: Bencivenga, M., Granetti, B., eds. *Atti del secondo congresso internazionale sul tartufo.* Spoleto, Italy, 24–27 November 1988. Comunità Montana dei Monti Martani e del Serano, Spoleto. Pp. 391–396.

Poitou, N., Olivier, J.M. 1990. La truffe et le cuivre. In: Bencivenga, M., Granetti, B., eds. *Atti del secondo congresso internazionale sul tartufo.* Spoleto, Italy, 24–27 November 1988. Comunità Montana dei Monti Martani e del Serano, Spoleto. Pp. 517–523.

Pollini, A. 2006. *Manuale di entomologia applicata.* Edagricole, Bologna.

Pomarico, M., Figliuolo, G., Rana, G.L., Alba, E. 2004. Assessment of biodiversity of truffle germplasm in Basilicata (Italy). In: *Proceedings of the XLVIII Italian Society of Agricultural Genetics–SIFV–SIGA Joint Meeting.* Lecce, Italy, 15–18 September 2004. Poster Abstract F.02.

Popham, P. 2004. Champion £28,000 truffle is laid to rest in Tuscany with full honours. *The Independent* (London). Available via www.news.independent.co.uk/europe/story.jsp?story=594578.

Potenza, L., Amicucci, A., Rossi, I., Palma, F., Bellis, R., Cardoni, P., Stocchi, V., De Bellis, R. 1994. Identification of *Tuber magnatum* Pico DNA markers by RAPD analysis. *Biotechnology Techniques* 8(2):93–98.

Powledge, T.M. 2004. The polymerase chain reaction. Available via www.faseb.org/prd/The%20Polymerase%20Chanin%20Reaction.pdf.

Pullar, W.A. 1962. *Soils and Agriculture of Gisborne Plains.* New Zealand Soil Bureau Bulletin no. 20.

Raggi Vivai. 2005. Tartufo. Available via www.raggivivai.it/prodotti/tartufo/sottomenu/listino.asp, www.raggivivai.it/prodotti/tartufo/sottomenu/aree_tartufigene/sardegna.asp, and www.raggivivai.it/prodotti/tartufo/sottomenu/aree_tartufigene/sicilia.asp.

Raglione, M., Owczarek, M. 2005. The soils of natural environments for growth of truffles in Italy. *Mycologia balcanica* 2:209–216.

Rambold, G., Agerer, R. 1997. DEEMY: the concept of a characterization and determination system for ectomycorrhizae. *Mycorrhiza* 7:113–116.

Ramírez Carrasco, R., Reyna Doménech, S., Suárez Olave, R. 2005. Current state and perspectives of truffle cultivation in Chile. In: *Proceedings of the Fourth International Workshop on Edible Mycorrhizal Mushrooms.* Murcia, Spain, 28 November–2 December 2005. Universidad de Murcia, Murcia. P. 106.

Ramsbottom, J. 1953. *Mushrooms and Toadstools.* Collins, London.

Ray, J. 1682. *Methodus plantarum nova, brevitatis and perspicuitatis causa synoptice in tabulis exhibita, cum notis generum tum summorum tum subalternorum characteristi-*

cis, observationibus nonnullis de seminibus plantarum and indice copioso. London.

Reale, A., Sorrentino, E., Maturo, L., Coppola, R. 2005. Sviluppo e applicazione di nuove tecnologie di conservazione dei tartufi. In: *Atti del convegno finale del progetto "Trattamento di prodotti freschi altamente deperibili per garantire qualità, sicurezza e salubrità".* Ariano Irpino, Italy, 15–19 March 2005. Universita degli Studi del Sannio, Ariano Irpino (AV), Italy. Pp. 132–137.

Rebière, J. 1967. *La truffe du Périgord.* Fanlac, Périgueux, France.

Redecker, D., Kodner, R., Graham, L.E. 2000a. Glomalean fungi from the Ordovician. *Science* 289:1920–1921.

Redecker, D., Morton, J.B., Bruns, T.D. 2000b. Ancestral lineages of arbuscular mycorrhizal fungi (Glomales). *Molecular Phylogenetics and Evolution* 14:276–284. Available via www.plantbio.berkeley.edu/~bruns/ftp/redecker2000a.pdf.

Reess, M. 1880. Über den Parasitismus von *Elaphomyces granulatus. Botanik zeitung* 38:729–733.

Renowden, G. 2005. *The Truffle Book.* Limestone Hills Publishing, Amberley, New Zealand.

Reuter, D.J., Robinson, J.B. 1986. *Plant Analysis.* Inkata Press, Melbourne.

Reyna, S. 1992. *La trufa.* Mundi-Prensa, Madrid.

Reyna, S. 2000. *Trufa, truficultura y selvicultura trufera.* Mundi-Prensa, Madrid.

Reyna, S., Boronat, J., Palomar, E. 2002a. Quality control of plants mycorrhized with *Tuber melanosporum* Vitt. In: Hall, I.R., Wang, Y., Danell, E., Zambonelli, A., eds. *Edible Mycorrhizal Mushrooms and Their Cultivation: Proceedings of the Second International Conference on Edible Mycorrhizal Mushrooms.* Christchurch, New Zealand, 3–5 July 2001. CD-ROM. New Zealand Institute for Crop and Food Research Limited, Christchurch.

Reyna, S., De Miguel, A., Palanzón, C., Hernández, A. 2005. Spanish trufficulture. In: *Proceedings of the Fourth International Workshop on Edible Mycorrhizal Mushrooms.* Murcia, Spain, 28 November–2 December 2005. Universidad de Murcia, Murcia. P. 109.

Reyna, S., Rodriguez Barreal, J.A., Folch, L., Pérez-Badía, R., Domínguez, A., Saiz-De-Omecaña, J.A., Zazo, J. 2002b. Techniques for inoculating mature trees with *Tuber melanosporum* Vitt. In: Hall, I.R., Wang, Y., Danell, E., Zambonelli, A., eds. *Edible Mycorrhizal Mushrooms and Their Cultivation: Proceedings of the Second International Conference on Edible Mycorrhizal Mushrooms.* Christchurch, New Zealand, 3–5 July 2001. CD-ROM. New Zealand Institute for Crop and Food Research Limited, Christchurch.

Ricard, J.-M., Bergounoux, F., Callot, G., Chevalier, G., Olivier, J.-M., Pargney, J.C., Sourzat, P. 2003. *La truffe guide technique.* Centre technique interprofessionnel des fruits et légumes, Paris.

Ridley, M. 2000. *Genome.* Harper Collins, New York.

Rincón, A., Alvarez, I.F., Pera, J. 2001. Inoculation of containerized *Pinus pinea* L. seedlings with seven ectomycorrhizal fungi. *Mycorrhiza* 11:265–271.

Riousset, L., Riousset, G., Chevalier, G., Bardet, M.C. 2001. *Truffes d'Europe et de Chine.* INRA, Paris.

Robin Pepinieres. 2001. www.robinpepinieres.com.

Rocchia, J.M. 1992a. *De la truffe en général et de la rabasse en particular.* Barthelemy, Paris.

Rocchia, J.M. 1992b. *Truffles: The Black Diamonds and Other Kinds.* Barthelemy, Avignon.

Rossi, I., Bartolacci, B., Potenza, L., Bertini, L., Barbieri, E., Stocchi, V. 2000. Identification of white truffle species using RAPD markers. *Plant and Soil* 219:127–133.

Rouch, P., Vercesi, A. 1990. La production en France de plants mycorhizés par la truffe selon le procédé INRA-ANVAR. In: Bencivenga, M., Granetti, B., eds. *Atti del secondo congresso internazionale sul tartufo.* Spoleto, Italy, 24–27 November 1988. Comunità Montana dei Monti Martani e del Serano, Spoleto. Pp. 483–493.

Rovesti, L., Viccinelli, R., Barbarossa, B. 1996. Biological control of sciarid flies. *Bulletin of the International Organization for Biological and Integrated Control of Noxious Animals and Plants* 19(9):20–23.

Royce, D.J. 1996. Specialty mushrooms. In: J. Janick, ed. *Progress in New Crops.* American Society for Horticultural Science Press, Arlington, Va. Pp. 464–475. Available via www.hort.purdue.edu/newcrop/proceedings1996/V3-464.html.

Rubini, A., Paolocci, F., Granetti, B., Arcioni, S. 1998. Single step molecular characterization of morphologically similar black truffle species. *FEMS Microbiology Letters* 164(1):7–12.

Rubini, A., Paolocci, F., Granetti, B., Arcioni, S. 2001. Morphological characterization of molecular-typed *Tuber magnatum* ectomycorrhizae. *Mycorrhiza* 11:179–185.

Rubini, A., Paolocci, F., Riccioni, C., Vendramin, G.G., Arcioni, S. 2005. Genetic and phylogeographic structures of the symbiotic fungus *Tuber magnatum. Applied and Environmental Microbiology* 71:6584–6589.

Saggs, H.W.F., Fairfield, H.N. 1965. Everyday life in Babylonia and Assyria. Assyrian International News Agency. Available via www.aina.org/books/eliba/eliba.pdf.

Sainte Alvere. 2005. Interprofessional agreement on fresh truffles. Available via www.sainte-alvere.com/uk_accord.asp.

Sanders, M. 2003. *From Here, You Can't See Paris: Seasons of a French Village.* Bantam, London.

Sbrana, C., Agnolucci, M., Bedini, S., Lepara, A., Toffanin, A., Giovannetti, M., Nuti, M.P. 2002. Diversity of culturable bacterial populations associated to *Tuber borchii* ectomycorrhizas and their activity on *T. borchii* mycelial growth. *FEMS Microbiology Letters* 211:195–201.

Sbrana, C., Bagnoli, G., Bedini, S., Filippi, C., Giovannetti, M., Nuti, M.P. 2000. Adhesion to hyphal matrix and antifungal activity of *Pseudomonas* strains isolated from *Tuber borchii* ascocarps. *Canadian Journal of Microbiology* 46:259–268.

Schrey, S.D., Schellhammer, M., Ecke, M., Hampp, R., Tarkka, M.T. 2005. Mycorrhiza helper bacterium *Streptomyces* AcH 505 induces differential gene expression in the ectomycorrhizal fungus *Amanita muscaria. New Phytologist* 168:205–216.

Schwendener, S. 1869. Die Flechten als Parasiten der Algen. *Verhandlungen der Schweizerischen Naturforschenden Gesellschaft Basel* 5:527–550.

Science in Africa. 2002. Threats to the environment posed by war in Iraq. Available via www.scienceinafrica.co.za/2003/march/war.htm.

References

Scortichini, M. 2002. Bacterial canker and decline of European hazelnut. *Plant Disease* July: 704–709. Available via www.apsnet.org/pd/pdfs/2002/0430-01F.pdf.

Scottish Executive Environment and Rural Affairs Department. 2005. *Phytophthora ramorum* and *kernoviae*. Available via www.scotland.gov.uk/Topics/Agriculture/ plant/17937/Phytophtras/SEERAD2004survey.

Séjalon Delmas, N., Roux, C., Martins, M., Kulifaj, M., Becard, G., Dargent, R. 2000. Molecular tools for the identification of *Tuber melanosporum* in agroindustry. *Journal of Agricultural and Food Chemistry* 48(6):2608–2613.

Selosse, M.A., Faccio, A., Scappaticci, G., Bonfante, P. 2004. Chlorophyllous and achlorophyllous specimens of *Epipactis microphilla* (Neottieae, Orchidaceae) are associated with ectomycorrhizal septomycetes, including truffles. *Microbiology Ecology* 47:415–442.

Senesi, E. 1990. Esperienze sulla conservazione e valorizzazione del tartufo bianco pregiato (*Tuber magnatum* Pico). In: Bencivenga, M., Granetti, B., eds. *Atti del secondo congresso internazionale sul tartufo*. Spoleto, Italy, 24–27 November 1988. Comunità Montana dei Monti Martani e del Serano, Spoleto. Pp. 611–619.

Setälä, H. 2000. Reciprocal interactions between Scots pine and soil food web structure in the presence and absence of ectomycorrhiza. *Oecologia* 125:109–118.

Shaw, T.M., Dighton, J., Sanders, F.E. 1995. Interactions between ectomycorrhizal and saprotrophic fungi on agar and in association with seedlings of lodgepole pine (*Pinus contorta*). *Mycological Research* 99:159–165.

Shorrocks, V.M. 1997. The occurrence and correction of boron deficiency. *Plant and Soil* 193:121–148.

Signorini, D., Valli, O. 1990. *Il tartufo*. Ottaviano-Mistral, Sommacampagna.

Silurian Software. 2006. The Silurian in Shropshire. Available via www.silurian. com/geology/salop.htm.

Silverstone, P. 2007. An interview with Peter Mayle. Available via www.restaurant-report.com/departments/w_petermayle.html.

Singer, R. 1961. *Mushrooms and Truffles*. Hill, London.

Singer, R., Harris, B. 1987. *Mushrooms and Truffles: Botany, Cultivation and Utilization*. Koeltz, Koenigstein, Germany.

Sisti, D., Giomaro, G., Cecchini, M., Faccio, A., Novero, M., Bonfante, P. 2003. Two genetically related strains of *Tuber borchii* produce *Tilia* mycorrhizas with different morphological traits. *Mycorrhiza* 13:107–115.

Sisti, D., Giomaro, G., Rossi, I., Ceccaroli, P., Citterio, B., Stocchi, V., Zambonelli, A., Benedetti, P.A. 1998. *In vitro* mycorrhizal synthesis of micropropagated *Tilia platyphyllos* Scop. plantlets with *Tuber borchii* Vittad. mycelium in pure culture. *Acta Horticulturae* 3:457–460.

Smith, S.E., Read, D.J. 1997. *Mycorrhizal Symbiosis*. Academic Press, London.

Sobin, G. 2000. *The Fly-Truffler*. Bloomsbury, London.

Solomon, J.D., McCracken, F.I., Anderson, R.L., Lewis, R. Jr., Oliveria, F.L., Filer, T.H., Barry, P.J. 2006. Oak pests: a guide to major insects, diseases, air pollution and chemical injury. U.S. Department of Agriculture Forest Service. Available via www.fs.fed.us/r8/foresthealth/pubs/oakpests/diseases_index.html.

Song, M.-S. 2005. Taxonomic and molecular systematic studies on *Tuber* in China.

Ph.D. diss., Systematic Mycology and Lichenology Laboratory, Institute of Micro-biology, Chinese Academy of Sciences, Beijing.

Song, M.-S., Cao, J.-Z., Yao, Y.-J. 2005. Occurrence of *Tuber aestivum* in China. *Mycotaxon* 91:75–80.

Sourzat, P. 1981. *Guide practique de trufficulture*. Délégation Départementale des Services d'Agronomie du Lot, Cahors–Le Montat.

Sourzat, P. 1989. *Guide practique de trufficulture*. 2nd ed. Délégation Départementale des Services d'Agronomie du Lot, Cahors–Le Montat.

Sourzat, P. 2001. Evolutions technologiques et ecologiques de la trufficulture en France et notaniment dans le Department du Lot. In: *Proceedings of the Fifth International Congress on the Science and Cultivation of Truffles*. Aix-en-Provence, France, 3–6 March 1999. Federation Française des Trufficulteurs, Paris. Pp. 419–424.

Sourzat, P. 2002. *Guide practique de trufficulture*. 4th ed. Station d'experimentation sur la truffe, Lycée professionnel agricole et viticole de Cahors–Le Montat.

Sourzat, P., Dubiau, J.M. 2001. Les contaminations par *Tuber brumale* dans les plantations truffieres: observations, experimentations et experimentations et strategie de protection en culture: exemple d'une plantation truffiere a miers, dans le Lot (France). In: *Proceedings of the Fifth International Congress on the Science and Cultivation of Truffles*. Aix-en-Provence, France, 3–6 March 1999. Federation Française des Trufficulteurs, Paris. Pp. 466–468.

Sourzat, P., Kulifaj, M., Montat, C. 1993. *Résultats techniques sur la trufficulture à partir d'expérimentations conduits dans le Lot entre 1985 et 1992*. Station d'Experimentations sur la Truffe/GIS Truffe, Le Montat.

Sourzat, P., Muratet, G., Schneider, J.P. 1990a. Observations sur la statut mycorhizien de jeunes arbres truffiers dans un essai de désinfection du sol au bromure de méthyle. In: Bencivenga, M., Granetti, B., eds. *Atti del secondo congresso internazionale sul tartufo*. Spoleto, Italy, 24–27 November 1988. Comunità Montana dei Monti Martani e del Serano, Spoleto. Pp. 283–288.

Sourzat, P., Muratet, G., Schneider, J.P. 1990b. La maîtrise des techniques de production truffière par l'experimentation au Lycée d'Enseignement Professionnel Agricole de Cahors-Le Montat. In: Bencivenga, M., Granetti, B., eds. *Atti del secondo congresso internazionale sul tartufo*. Spoleto, Italy, 24–27 November 1988. Comunità Montana dei Monti Martani e del Serano, Spoleto. Pp. 467–473.

Spencer, R., Randall, C. 2006. Chinese truffles put the wind up French traders. *Daily Telegraph*, 26 January. Available via www.telegraph.co.uk/news/main.jhtml?xml=/news/2006/01/26/wtruf26.xml&sSheet=/news/ 2006/01/26/ixworld.html.

Splivallo, R. 2004. Truffle volatiles and exudates: identification and biological effects. Ph.D. diss. progress report, University of Torino, Torino. Available via www.bioveg.unito.it/italiano/reldot04/splivallo.doc.

Splivallo, R., Bossi, S., Maffei, M., Bonfante, P. 2007. Discrimination of truffle fruiting body versus mycelial aromas by stir bar sorptive extraction. *Phytochemistry* (2007), doi:10.1016/j.phytochem.2007.03.030.

Splivallo, R., Novero, M., Bossi, S., Bonfante, P. 2007. Truffle volatiles inhibit

growth and induce an oxidative burst in *Arabidopsis thaliana*. *New Phytologist* (2007) doi:10.1111/j.1469-8137.2007.02141.x.

Stamets, P. 2000. *Growing Gourmet and Medicinal Mushrooms*. 3rd ed. Ten Speed Press, Toronto.

Steinmann, S.H. 1998. Gender, pastoralism, and intensification: changing environmental resource use in Morocco. In: Coppock, J., Miller, J.A., Albert, J., Bernhardsson, M., Kenna, R. eds. *Transformations of Middle Eastern Natural Environments: Legacies and Lessons*. Yale School of Forestry and Environmental Studies Bulletin no. 103. Pp. 81–107. Available via http://environment.yale.edu/documents/downloads/0-9/103steinmann.pdf.

Stringer, A. 2004. *Boletus edulis* in New Zealand: its genetic affinities and history. MSc. thesis, University of Otago, Dunedin.

Sussex Biodiversity Partnership. 2001. The biodiversity action plan for East and West Sussex, Brighton and Hove. Habitat action plan for Sussex: Sussex woodlands. Available via www.biodiversitysussex.org/woodland.htm.

Switzer, S. 1727. *The Practical Kitchen Gardener*. Woodward, London.

Szemere, L. 1965. *Die unterirdischen Pilze des Karpatenbeckens: Fungi Hypogaei territorii Carpato-Pannonici*. Akadémiai Kiado, Budapest.

Taber, R.A. 1990. Observations on the origin of the Texas truffle, *Tuber texense*. In: Bencivenga, M., Granetti, B., eds. *Atti del secondo congresso internazionale sul tartufo*. Spoleto, Italy, 24–27 November 1988. Comunità Montana dei Monti Martani e del Serano, Spoleto. Pp. 331–336.

Talou, T., Delmas, M., Gaset, A. 1988. Black truffle hunting: use of gas detectors. *Transactions of the British Mycological Society* 91:337–338.

Tanesaka, E., Masuda, H., Kinugawa, K. 1993. Wood degrading ability of basidiomycetes that are wood decomposers, litter decomposers, or mycorrhizal symbionts. *Mycologia* 85:347–354.

Tanfulli, M., Giovanotti, E., Donnini, D., Baciarelli Falini, L. 2001. Analisi della micorrizazione in tartufaie coltivate di *Tuber aestivum* Vittad. e *T. borchii* Vittad. impiantate da oltre 12 anni in ambienti pedoclimatici diversi. In: *Proceedings of the Fifth International Congress on the Science and Cultivation of Truffles*. Aix-en-Provence, France, 3–6 March 1999. Federation Française des Trufficulteurs, Paris. Pp. 480–484.

Tannahill, R. 1995. *Food in History*. Three Rivers Press, New York.

Tao, K., Liu, B., Zhang, D. 1989. A new species of *Tuber* from China. *Journal of Shanxi University* 12:215–218.

Taylor, D.L., Bruns, T.D. 1999. Community structure of ectomycorrhizal fungi in a *Pinus muricata* forest: minimal overlap between the mature forest and resistant propagule communities. *Molecular Ecology* 8:1837–1850.

Taylor, F.W., Thamage, D.M., Baker, N., Roth-Bejerano, N., Kagan-Zur, V. 1995. Notes on the Kalahari Desert truffle, *Terfezia pfeilii*. *Mycological Research* 99:874–878.

Taylor, P. 2005. Giant truffle a feast of patience. *The Australian* 16 June.

Tedersoo, L., Hansen, K., Perry, B.A., Kjøller, R. 2006. Molecular and morphological diversity of pezizalean ectomycorrhiza. *New Phytologist* 170:581–596.

Thompson, S. 2006. FRST briefs the minister. Royal Society Alert no. 409. Available via www.rsnz.org/news/sciencealert.php.

Tibiletti, E., Zambonelli, A. 1999. *I tartufi della provincia di Forlì-Cesena*. Pàtron, Bologna.

Times Online, Press Association. 2004. Giant £28,000 truffle rots in chef's safe. *Times* (London), 8 December. Available via www.timesonline.co.uk/tol/news/uk/article400621.ece.

Tirillini, B., Verdelli, G., Paolocci, F., Ciccioli, P., Frattoni, M. 2000. The volatile organic compounds from the mycelium of *Tuber borchii* Vitt. *Phytochemistry* 55:983–985.

Tocci, A. 1982. Tartuficoltura: elementi per una razionale coltivazione. *Agricoltura e ricerca* 18:48–55.

Tocci, A. 1985. Ecologia del *Tuber magnatum* Pico nell'Italia centrale. *Annali dell'istituto sperimentale per la selvicoltura Arezzo* 16:425–541.

Tocci, A. 2001. Il tartufo ha nobilitato la cucina Italiana. In: *Proceedings of the Fifth International Congress on the Science and Cultivation of Truffles*. Aix-en-Provence, France, 3–6 March 1999. Federation Française des Trufficulteurs, Paris. Pp. 522–524.

Tocci, A., Gregori, G., Chevalier, G. 1985. Produzione di piantine tartufagene (*Tuber magnatum* Pico): sintesi micorrizica col sistema dell'innesto radicale. *Italia forestale e montana* 3:143–153.

Tocci, A., Gregori, G., Denci, L. 1986. *I tartufi della legge quadro nazionale*. Ministero dell'Agricoltura e delle foreste, Republica Italiana, Corpo forestale dello stato, Rome.

Tous, J., Romero, A., Plana, J., Sentis, X., Ferrán, J. 2005. Effect of nitrogen, boron and iron fertilization on yield and nut quality of 'negret' hazelnut trees. In: Tous, J., Rovira, M., Romero, A. eds. VI International Congress on Hazelnut. *ISHS Acta Horticulturae* 686:277–280.

Toy, S., Kapner, F., Westley, A. 1990. Pity the truffler . . . his troubles are mushrooming. *Business Week*, 17 December, 36.

Trappe, J.M. 1977. Selection of fungi for ectomycorrhizal inoculation in nurseries. *Annual Review of Phytopathology* 15:203–222.

Trappe, J.M. 1979. The orders families and genera of hypogeous Ascomycotina (truffles and their relatives). *Mycotaxon* 9:297–340.

Trappe, J.M. 1985. Translation of B. Frank. 1885. On the root symbiosis-depending-nutrition through hypogeous fungi of certain trees. In: Molina R., ed. *Proceedings of the Sixth North American Conference on Mycorrhizae*. Bend, Oregon, June 1984. Forest Research Laboratory, Corvallis, Ore. Pp. 25–29.

Trappe, J.M. 1990. Use of truffles and false-truffles around the world. In: Bencivenga, M., Granetti, B., eds. *Atti del secondo congresso internazionale sul tartufo*. Spoleto, Italy, 24–27 November 1988. Comunità Montana dei Monti Martani e del Serano, Spoleto. Pp. 19–30.

Trappe, J.M. 2004. A. B. Frank and mycorrhizae: the challenge to evolutionary and ecologic theory. *Mycorrhiza* 15:277–281.

Trappe, J.M., Castellano, M.A. 2005. Keys to the genera of truffles (Ascomycetes). Available via www.natruffling.org/ascokey.htm.

Trappe, J.M., Claridge, A.W., Claridge, D.L. Desert truffles of the Australian outback and African Kalahari: ecology, ethnomycology and taxonomy. In prep.

Trappe, J.M., Jumpponen, A.M., Cazares, E. 1996. NATS truffle and truffle-like fungi. 5. *Tuber lyonii* (= *T. texense*), with a key to the spiny-spored *Tuber* species groups. *Mycotaxon* 60:365–372.

Trappe, J.M., Maser, C. 1977. Ectomycorrhizal fungi: interactions of mushrooms and truffles with beasts and trees. In: Walters, T., ed. *Mushrooms and Man: An Interdisciplinary Approach to Mycology.* U.S. Department of Agriculture Forest Service, Pacific Northwest Station, Corvallis, Ore. Pp. 165–179.

Trappe, J.M., Molina, R. 1986. Taxonomy and genetics of mycorrhizal fungi: their interactions and relevance. In: Gianinazzi-Pearson, V., Gianinazzi, S., eds. *Mycorrhizae: Physiology and Genetics.* INRA, Paris. Pp. 133–146.

Treehelp.com. 2006. Oak trees: other problems—powdery mildew. Available via www.treehelp.com/trees/oak/oak-diseases-powdery-mildew.asp.

Treessentials. 2006. Clipper Grow Tube. Available via www.growtubes.com.

Trevor, E., Yu, J.-C., Egger, K.N., Peterson, R.L. 2001. Ectendomycorrhizal associations: characteristics and functions. *Mycorrhiza* 11:167–177.

Troisgros, J., Troisgros, P. 1978. *The Nouvelle Cuisine of Jean and Pierre Troisgros.* Translated by R.W. Smoler. Morrow, New York.

Truffel.com. 2005. www.truffel.com/pages/luoQ.asp.

Truffle UK. 2006. Summer truffle in the UK. Available via www.truffle-uk.co.uk/home_uk_summer_truffle.php and www.truffle-uk.co.uk/news_press_cuttings.php.

Tuber.it. 2006. La borsa del tartufo. Available via www.tuber.it/borsa.php.

Tubex Treeshelters. 2006. www.tubex.com/index.php?page=43.

Tulasne, L.R. 1851. *Fungi hypogaei.* Paris.

Turner, P.C., McLennan, A.G., Bates, A.D., White, M.R.H. 2001. *Molecular Biology.* Bios, Oxford.

United Nations. 2004. UNECE Recommendation FFV-53 concerning the marketing and commercial quality control of fresh truffles (*Tuber*). Available via www.unece.org/trade/agr/meetings/ge.01/document/2004_25_a08.pdf.

University of California Cooperative Extension. 1998. Orchard facts: precautions for using potassium chloride for correction of potassium deficiency. Available via www.ceglenn.ucdavis.edu/newsletterfiles/Orchard_Facts592.PDF.

University of Canterbury. 2006. Weather charts and weather stations. Department of Geography, University of Canterbury, Christchurch, New Zealand. Available via www.geog.canterbury.ac.nz/weather/rooftop/html/current.html and www.geog.canterbury.ac.nz/weather/weather.htm.

Urbani. 2007. www.urbanitartufi.it/.

U.S. Department of Agriculture. 1997. BT (*Bacillus thuringiensis*) toxin resources. Available via www.nal.usda.gov/bic/BTTOX/bttoxin.htm.

Van der Walt, P., Le Riche, E. 1999. *The Kalahari and Its Plants.* Info Naturae, Pretoria.

Verlhac, A., Girard, M., Pallier, R. 1989. Truffe: fertilisation organique en truffière productrice irriguée. *Infos* 56:22–26.

Vilgaly, R. 2007. Duke mycology: Conserved primer sequences for PCR amplification and sequencing from nuclear ribosomal RNA. Department of Biology, Duke
University. Available via www.biology.duke.edu/fungi/mycolab/primers.htm.

Vinay, M., Pirazzi, R. 2001. Realtà ed esigenze per la coltivazione di *T. borchii* Vittad. e *T. aestivum* Vittad. nell'Italia centrale. In: *Proceedings of the Fifth International Congress on the Science and Cultivation of Truffles*. Aix-en-Provence, France,
3–6 March 1999. Federation Française des Trufficulteurs, Paris. Pp. 425–430.

Vitosh, M.L., Warncke, D.D., Lucas, R.E. 1994. Secondary and micronutrients for
vegetables and field crops. Michigan State University Extension Bulletin E-486.
Available via www.web1.msue.msu.edu/msue/imp/modf1/05209709.html.

Vittadini, C. 1837. *Monographia tuberacearum*. Mediolani.

Vogt, K.A., Bloomfield, J., Ammirati, J.F., Ammirati, S.R. 1992. Sporocarp production by Basidiomycetes, with emphasis on forest ecosystems. In: Carroll, G.C.,
Wicklow, D.T. eds. *The Fungal Community: Its Organization and Role in the Ecosystem*. 2nd ed. Marcel Dekker, New York. Pp. 563–581.

Wallén, C.C., ed. 1970. *World Survey of Climatology*. Vol. 5, *Climates of Northern and
Western Europe*. Elsevier, Amsterdam.

Wallén, C.C., ed. 1977. *World Survey of Climatology*. Vol. 6, *Climates of Central and
Southern Europe*. Elsevier, Amsterdam.

Wang, X., Liu, P., Yu, F. 2004. *Colour Atlas of Wild Commercial Mushrooms in China*.
Yunnan Science and Technology Press, Kunming, China.

Wang, Y. 1990. First report of a study on *Tuber* species from China. In: Bencivenga,
M., Granetti, B., eds. *Atti del secondo congresso internazionale sul tartufo*. Spoleto,
Italy, 24–27 November 1988. Comunità Montana dei Monti Martani e del Serano,
Spoleto. Pp. 45–52.

Wang, Y., Hall, I.R. 2001. *Tuber sinense* and other *Tuber* species from south-west
China. In: Courvoisier, M., ed. *Science et culture de la truffe et des autres champignons
hypogés comestibles*. Fédération Française des Trufficulteurs, Paris. Pp. 115–116.

Wang, Y., Hall, I.R. 2004. Edible ectomycorrhizal mushrooms: challenges and
achievements. *Canadian Journal of Botany* 82:1063–1073.

Wang, Y., Hall, I.R., Evans, L. 1997. Ectomycorrhizal fungi with edible fruiting
bodies. 1. *Tricholoma matsutake* and allied fungi. *Economic Botany* 51:311–327.

Wang, Y., He, X.-Y. 2002. *Tuber huidongense* sp. nov. from China. *Mycotaxon*
83:191–194.

Wang, Y., Li, Z.-P. 1991. A new species of *Tuber* from China. *Acta Mycologia Sinica*
10:263–265.

Wang, Y., Moreno, G., Riousset, L.J., Manjon, J.L., Riousset, G., Fourre, G., Di
Massimo, G., Garcia-Montero, L.G., Diez, J. 1998. *Tuber pseudoexcavatum* sp. nov.,
a new species from China commercialised in Spain, France and Italy with additional comments on Chinese truffles. *Cryptogamie mycologie* 19:113–120.

Wang, Y., Tan, Z.M., Zhang, D.C., Murat, C., Jeandroz, S., Le Tacon, F. 2006a.
Phylogenetic and populational study of the *Tuber indicum* complex. *Mycological
Research* 110:1034–1045.

Wang, Y., Tan, Z.M., Zhang, D.C., Murat, C., Jeandroz, S., Le Tacon, F. 2006b.
Phylogenetic relationships between *Tuber pseudoexcavatum*, a Chinese truffle, and

other *Tuber* species based on parsimony and distance analysis of four different gene sequences. *FEMS Microbiology Letters* 259:269–281.

Warder, M. 2007. Plants in the Kalahari Desert. Available via www.abbott-infotech. co.za/kalahari%20desert%20pictures.html.

Watson, J.D., Berry, A. 2003. *DNA: The Secret of Life*. Knopf, New York.

Wedén, C., ed. 2004. Black truffles of Sweden: systematics, population studies, ecology and cultivation of *Tuber aestivum* syn. *T. uncinatum*. Acta Universitatis Upsaliensis: Comprehensive Summaries of Uppsala Dissertations from the Faculty of Science and Technology no. 1043. Available via http://publications.uu.se/abstract. xsql?dbid=4675.

Wedén, C., Chevalier, G., Danell, E. 2004a. *Tuber aestivum* (syn. *T. uncinatum*) biotopes and their history on Gotland, Sweden. *Mycological Research* 108:304–310.

Wedén, C., Danell, E. 2001. *Tuber aestivum* in Sweden. In: *Proceedings of the Fifth International Congress on the Science and Cultivation of Truffles*. Aix-en-Provence, France, 3–6 March 1999. Federation Française des Trufficulteurs, Paris. P. 247.

Wedén, C., Danell, E. 2002. Scandinavian black truffles: distribution and habitats. In: Hall, I.R., Wang, Y., Danell, E., Zambonelli, A., eds. *Edible Mycorrhizal Mushrooms and Their Cultivation: Proceedings of the Second International Conference on Edible Mycorrhizal Mushrooms*. Christchurch, New Zealand, 3–5 July 2001. CD-ROM. New Zealand Institute for Crop and Food Research Limited, Christchurch.

Wedén, C., Danell, E., Camacho, F.J., Backlund, A. 2004b. The population of the hypogeous fungus *Tuber aestivum* syn. *T. uncinatum* on the island of Gotland. *Mycorrhiza* 14:19–23.

Wedén, C., Danell, E., Tibell, L. 2005. Species recognition in the truffle genus *Tuber*: the synonyms *Tuber aestivum* and *Tuber uncinatum*. *Environmental Microbiology* 7:1535–1546.

Wedén, C., Pettersson, L., Danell, E. 2004c. Cultivation of the Burgundy truffle, *Tuber aestivum* syn. *T. uncinatum*, in Sweden. In: Wedén, C., ed. Black truffles of Sweden: systematics, population studies, ecology and cultivation of *Tuber aestivum* syn. *T. uncinatum*. Acta Universitatis Upsaliensis: Comprehensive Summaries of Uppsala Dissertations from the Faculty of Science and Technology no. 1043. Available via http://publications.uu.se/abstract.xsql?dbid=4675.

Wenkart, S., Roth-Bejerano, N., Mills, D., Kagan-Zur, V. 2001. Mycorrhizal associations between *Tuber melanosporum* mycelia and transformed roots of *Cistus incanus*. *Plant Cell Reports* 20:369–373.

Wessex Coppice Group. 2006. Background information. Available via www.coppice. org.uk/background.htm.

Wheeler, D.B. 2005. Oregon white truffles. Available via www.oregonwhitetruffles.com.

White, T.J., Bruns, T., Lee, S., Taylor, J. 1990. Amplification and direct sequencing of fungal ribosomal RNA genes for phylogenetics. In: Innis, M.A., Gelfand, D.H., Sninsky, J.J., White, T.J., eds. *PCR Protocols: A Guide to Methods and Applications*. Academic Press, San Diego. Pp. 315–322.

Wikipedia. 2006. Truffle: most expensive truffle. Available via http://en.wikipedia. org/wiki/White_truffle.

Willey, D. 2004. Record-breaking truffle goes home. Available via www.news.bbc. co.uk/1/hi/world/europe/4107083.stm.

Wood, A. 2006. Compendium of pesticide common names: fungicides. Available via www.alanwood.net/pesticides/index.html.

Wurzburger, N., Bledsoe, C.S. 2001. Comparison of ericoid and ectomycorrhizal colonization and ectomycorrhizal morphotypes in mixed conifer and pygmy forests on the northern California coast. *Canadian Journal of Botany* 79:1202–1210.

Yunnan Yunri Health Food Industry Development Co. Ltd. 2004. www. chinesetruffle.com.

Zacchi, L., Vaughan-Martini, A., Angelini, P. 2003. Yeast distribution in a trufflefield ecosystem. *Annals of Microbiology* 53:275–282.

Zambonelli, A. 1990. Confronto fra diversi metodi di conservazione dell'inoculo di *Tuber* spp. *Micologia Italiana* 19(3):25–29.

Zambonelli, A. 1993. Effetto dei trattamenti contro l'oidio della quercia sulle micorrize di *Tuber albidum* e *T. aestivum*. *Informatore fitopatologia* 43(2):59–62.

Zambonelli, A., Branzanti, M.B. 1984. Prove di micorrizazione del nocciolo con *T. aestivum* e *T. albidum*. *Micologia Italiana* 13:47–52.

Zambonelli, A., Branzanti, M.B. 1987. Competizione tra due funghi ectomicorrizici: *Tuber albidum* e *Laccaria laccata*. *Micologia Italiana* 16:159–164.

Zambonelli, A., Branzanti, M.B. 1989. Mycorrhizal synthesis of *Tuber albidum* Pico with *Castanea sativa* Mill. and *Alnus cordata* Loisel. *Agricultural Ecosystems and Environment* 28:563–567.

Zambonelli, A., Di Munno, R. 1992. *Indagine sulla possibilità di diffusione dei rimboschimenti con specie tartufigene: aspetti tecnico-colturali ed economici*. Ministero dell'Agricoltura e delle Foreste, Rome.

Zambonelli, A., Giunchedi, L., Poggi Pollini, C. 1993a. An enzyme-linked immunosorbent assay (ELISA) for the detection of *Tuber albidum* ectomycorrhiza. *Symbiosis* 15:71–76.

Zambonelli, A., Govi, G., Previati, A. 1989. Micorrizazione *in vitro* di piantine micropropagate di *Populus alba* con micelio di *Tuber albidum* in coltura pura. *Micologia Italiana* 18:105–111.

Zambonelli, A., Iotti, M. 2001. Effects of fungicides on *Tuber borchii* and *Hebeloma sinapizans* ectomycorrhizas. *Mycological Research* 105:611–614.

Zambonelli, A., Iotti, M. 2005. *Appennino Modenese terre da tartufo*. Giorgio Mondadori, Modena.

Zambonelli, A., Iotti, M. 2006. The pure culture of *Tuber* mycelia and their use in the cultivation of the truffles. In: Khabar, L., ed. *Le premier symposium sur les champignons hypogés du basin Méditerranéen*. Rabat, Morocco, 5–8 April 2004. Pp. 244–255.

Zambonelli, A., Iotti, M., Amicucci, A., Pisi, A. 1999. Caratterizzazione anatomo-morfologica delle micorrize di *Tuber maculatum* Vittad. su *Ostrya carpinifolia* Scop. *Micologia Italiana* 28:29–35.

Zambonelli, A., Iotti, M., Giomaro, G., Hall, I., Stocchi, V. 2002. *Tuber borchii* cultivation: an interesting perspective. In: Hall, I.R., Wang, Y., Danell, E., Zam-

bonelli, A., eds. *Edible Mycorrhizal Mushrooms and Their Cultivation: Proceedings of the Second International Conference on Edible Mycorrhizal Mushrooms*. Christchurch, New Zealand, 3–5 July 2001. CD-ROM. New Zealand Institute for Crop and Food Research Limited, Christchurch.

Zambonelli, A., Iotti, M., Rossi, I., Hall, I.R. 2000a. Interaction between *Tuber borchii* and other ectomycorrhizal fungi in a field plantation. *Mycological Research* 104:698–702.

Zambonelli, A., Iotti, M., Zinoni, F., Dallavalle, E., Hall, I.R. 2005. Effect of mulching on *Tuber uncinatum* ectomycorrhizas in an experimental truffière. *New Zealand Journal of Crop and Horticultural Science* 33:65–73.

Zambonelli, A., Penjor, D., Pisi, A. 1995a. Effetto del triadimefon sulle micorrize di *Tuber borchii* Vitt. e di *Hebeloma sinapizans* (Paulet) Gill. *Micologia Italiana* 24(3):65–73.

Zambonelli, A., Pisi, A., Tibiletti, A. 1997. Caratterizzazione anatomo-morfologica delle micorrize di *Tuber indicum* Cooke and Masee su *Pinus pinea* L. e *Quercus cerris* L. *Micologia Italiana* 26(1):29–36.

Zambonelli, A., Rivetti, C., Percudani, R., Ottonello, S. 2000b. TuberKey: a DELTA-based tool for the description and interactive identification of truffles. *Mycotaxon* 74:57–76. Available via www.truffle.org/tuberkey/tuberkey-english.html.

Zambonelli, A., Salomoni, S. 1993. Tartuficoltura, che fare. *Terra e vita* 41:38–39.

Zambonelli, A., Salomoni, S., Pisi, A. 1993b. Caratterizzazione anatomo-morfologica e micromorfologica delle micorrize di *Tuber* spp. su *Quercus pubescens* Willd. *Micologica Italiana* 22(3):73–90.

Zambonelli, A., Salomoni, S., Pisi, A. 1995b. Caratterizzazione anatomo-morfologica e micromorfologica delle micorrize di *Tuber borchii* Vitt., *Tuber aestivum* Vitt., *Tuber mesentericum* Vitt., *Tuber brumale* Vitt., *Tuber melanosporum* Vitt. su *Pinus pinea*. L. *Micologia Italiana* 24(2):119–137.

Zekri, M., Obreza, T.A. 2006. Micronutrient deficiencies in citrus: boron, copper, and molybdenum. University of Florida Extension Report SL203. Available via www.edis.ifas.ufl.edu/pdffiles/SS/SS42200.pdf.

Zhang, B.C., Minter, D.W. 1988. *Tuber himalayense* sp. nov. with notes on Himalayan truffles. *Transactions of the British Mycological Society* 91:953–957.

Zhang, L., Yang, Z.L., Song, D.S. 2005. A phylogenetic study of commercial Chinese truffles and their allies: taxonomic implications. *FEMS Microbiology Letters* 245:85–92.

Zigante tartufi. 2005. www.zigantetartufi.com/novo/index.php.

Zuccherelli, G. 1990. Moltiplicazione in vitro di cinque varietà di nocciolo e loro micorrizazione con *Tuber melanosporum*. *L'informatore agrario* 41:51–55.

Zuccherelli, G., Zuccherelli, S., Capaccio, V. 1992. Production *in vitro* of two hazelnut varieties and results about inoculum of *Tuber magnatum* Pico on a mass scale. Paper presented at the Third International Congress on Hazelnut, Alba, Italy. *Acta Horticulturae* 351:371–380.

INDEX

Abies, 233
Abies alba, 234, 237
Acacia erioloba, 118, 235, 239
Acacia hebeclada, 235, 237
Acer, 231
Acer negundo, 231
acidic soil, 115, 118, 135, 139, 179, 180
Aesculus, 231
Agathis, 232
Agrobacterium rhizogenes, 51
Agropyron repens, 172
Ailanthus altissima, 232
alkaline soil, 115, 118
Alnus, 112, 233, 234, 236
Alnus cordata, 112, 234, 236
Amanita muscaria, 41
Ampelomyces quisqualis, 186
androstenol, 204
angle droix (AD) fungus, 94, 95
Anisogramma anomala, 156, 186
Aralia, 231
arbuscular mycorrhiza, 36–38, 51, 124–126, 133, 139, 140, 231, 233
Arbutus, 233
Archips rosanus, 191
Armillaria, 187, 195
Ascospore, 28
ascus (pl. asci), 28, 45–47, 59–89

Asiatic truffles, 17, 67, 69, 153, 154, 223, 224, 226

Bacillus thuringensis, 192
Bagnoli truffle. *See Tuber mesentericum*
Balsamia, 55, 57, 86, 87
Balsamia vulgaris Vittad., 57, 86, 87
Bambusa, 231
beetle, 38, 190, 192, 193
Berberis, 231
Betula, 106, 156, 233
Betula verrucosa, 234, 236
Bianchetto. *See Tuber borchii*
black truffle. *See Tuber aestivum, Tuber brumale, Tuber macrosporum, Tuber melanosporum, Tuber mesentericum*
Boletus edulis, 38, 174
boron, 128, 129, 135, 179, 182, 246, 249
Botryosphaeria, 187
Boudier's terfez. *See Terfezia boudieri*
brûlé, 21, 70, 119, 144, 148, 167–177, 202
Burgundy truffle. *See Tuber aestivum*
Buxus, 231

calcium carbonate. *See* limestone
Camellia, 231
carbon, organic, 127, 246
Carica papaya, 232
Carpinus, 106, 233, 234, 236

Carpinus betulus, 106, 234, 236

Carpodetus serratus, 232

Carya, 80, 233

Carya illinoiensis, 80

Carya ovata, 80

Castanea, 156

Castanea sativa, 156

Casuarina, 233

Catalpa, 231

Cedrus, 118, 233

Cedrus atlantica, 234, 238

Cedrus deodara, 234, 238

Celtis, 231

Cerambyx cerdo, 189

Ceratocystis fagacearum, 159

Cercis canadensis, 233

Chamaecyparis lawsoniana, 232

Chinese truffles, 67–69, 223, 250

Choiromyces, 55, 204

Choiromyces echinulatus. See *Eremiomyces echinulatus*

Choiromyces meandriformis Vittad., 57, 73, 86, 87, 223

Choiromyces venosus. See *Choiromyces meandriformis*

Cistus, 51, 118, 233

Cistus albidus, 235, 238

Cistus incanus, 51, 238

Cistus monspeliensis, 51, 238

Cistus salviaefolius, 51, 238

Citrullus vulgaris, 119

classification, 27, 28, 53, 55

clay, 70, 105, 109, 117, 118, 126, 127, 135, 162, 176

climate, 17–20, 52, 98, 103, 110, 112, 115, 118, 121, 124, 174, 177, 194, 212, 242–245, 247

clone, 157

competing ectomycorrhizal fungi, 45, 47, 91, 100, 121–126, 131–134, 139, 142, 152–155, 160–163, 175, 177, 185, 197, 213, 214, 231, 248

compost, 48, 177

copper, 128, 129, 135, 179–183, 186–188

Coprosma robusta, 232

Cornus, 231

Corokia buddleoides, 232

Corylobium avellanae, 191

Corylus, 146, 233

Corylus americana, 186

Corylus avellana, 54, 106, 112, 156, 186, 216, 234, 237

Corylus colurna, 157, 234, 237

Corylus heterophylla, 157, 234, 237

Cossus cossus, 189

Costelytra zealandica, 192

Crataegus, 80, 233

Cryptomeria japonica, 231

cultivation, 21, 32–35, 41–52, 98–119, 121–162, 163–197

×*Cupressocyparis leylandii*, 232

Cupressus macrocarpa, 232

Cycad, 38, 124, 232

Cystidia, 37–40, 91–95

Cytospora corylicola, 187

Delastria, 55, 85

Delastria rosea Tul. & Tul., 241

desert truffles. See *Kalaharituber, Terfezia, Tirmania*

dimethyl sulphide, 57

Diospyros, 232

diseases, 30, 35, 85, 130, 150, 155, 156, 159, 163, 177–189, 192, 193, 229, 248

distribution of truffles, 98–120, 219, 241

DNA, 57, 76, 96, 97, 115, 229

Dodonaea viscose, 231

dogs, 20, 78, 86, 88, 99, 194, 200, 204–213

dolomite, 48, 135, 137, 179

drainage, 126, 127, 135, 141, 177, 187

drought, 148–152, 161, 168, 175–177, 182, 184, 217, 247

ectomycorrhiza, 36–41

EDDHA (ethylenediaminedi[o-hydroxyphenylacetyl] acid), 180

EDTA (ethylenediaminetetraacetic acid), 249

Elaphomyces, 55

elevation, 70, 81, 102, 103, 110, 142, 165, 242, 248

ELISA (enzyme linked immunosorbent assay), 189

Eremiomyces echinulatus (Trappe & Marasas) Trappe & Kagan-Zur, 15, 83, 118

Erysiphe trina, 185

Erythrina indica, 231

Eucalyptus, 233

Eulecanium tiliae, 191

Euonymus, 231

Fagus, 118, 233

Fagus sylvatica, 106, 187, 234, 237

false truffles, 88, 90

family, 54

fencing, 126, 137

fertilizers, 45, 48, 137, 177, 182, 249

Festuca ovina, 168

Ficus carica, 231

field inoculations, 50

folklore, 19, 148, 164

foliar analysis, 137, 178

fraud, 212, 213, 226

Fraxinus, 231

frost, 157, 161, 184, 188

Fuchsia, 231

Fumana procumbens, 235, 238

fungicides, 181, 185, 188

Gautieria, 90

Genea, 55, 88, 89

genus, 54

Ginkgo biloba, 231

gleba, 58–88

glufosinate-ammonium, 172, 194

glyphosate, 133, 172, 173

grass, 53, 56, 103, 119, 124, 133, 160, 172, 192, 203

grazing, 43

Griselinia, 231

Gypsophila, 160

habitats, 73, 98–120, 141, 198, 199, 201

Hartig net, 39, 119

harvesting, 42, 198–211, 241, 250

Hebeloma, 122, 185

Hebeloma radicosum, 229

Helianthemum, 52, 118, 162, 233

Helianthemum almeriense, 52, 148, 235, 239

Helianthemum guttatum, 235, 239

Helianthemum hirtum, 235, 239

Helianthemum ledifolium, 235, 239

Helianthemum lippii, 235, 239

Helianthemum marifolium, 235, 239

Helianthemum ovatum, 52, 235, 239

Helianthemum salicifolium, 235, 239

Helvella, 55, 56

Helvella crispa, 56

herbarium, 59, 81

herbicides, 133, 137, 157, 172, 173, 184

Heteronychus arator, 190

Hibiscus rosa-sinensis, 231

Hieracium pilosella, 168

Hoheria populnea, 232

host plants, 38, 41, 48, 80–83, 112, 118, 156–160, 234–239

Hydnangium, 90

Hydnobolites, 55

Hydnotria, 55

hydrated lime, 135, 214

Hymenogaster, 90

Hypoxylon mediterraneum, 187

identifying truffles, 57–97

Ilex, 231

infection, 11, 40–53, 92, 114, 122–124, 152–160, 163, 173, 183–189

inoculation, inoculum (pl. inocula), 41–52, 114, 152–159, 213, 228

insecticide, 189, 190, 192

internal transcribed spacer (ITS), 96, 97

International Code of Botanical Nomenclature, 54

iron, 112, 125, 128–131, 178–181, 246, 249

irrigation, 126, 127, 142, 148–152, 160, 163, 165, 173–177, 182, 183, 190, 215, 248

Italian black truffle of Fragno. *See Tuber aestivum*
Italian white truffle. *See Tuber magnatum*
Juglans, 233
Juniperus, 232

Kalahari truffle. *See Kalaharituber*
Kalaharituber, 15, 83, 86, 118,
Kalaharituber pfeilii (Henn.) Trappe & Kagan-Zur, 84, 85, 119, 241
Kamaa. *See Tirmania nivea*
Khulasi. *See Tirmania pinoyi*
kingdom, 54
Koelreuteria elegans, 232
Kunzea ericoides, 233

Lactarius deliciosus, 40
lapse rate, 102
Larix, 118, 233, 234, 238
laws and regulations, 21, 155, 198, 209, 211, 213, 224
Leptospermum, 38, 233
Leptospermum scoparium, 233
Leucangium, 81, 82
Leucangium carthusianum (Tul. & Tul.) Paol., 81, 82
Leucogaster, 90
Leucoma salicis, 192
Ligustrum, 232
limestone, 48, 99–118, 126–136, 143, 165, 175, 177, 180, 214
Linnaeus, Carl (Carl von Linné), 54
Liodes cinnamomea, 193
Liriodendron, 232
Loculotuber gennadii (Chatin) Trappe, Parladé & I.F. Alvarez, 85, 88, 241
Lymantria dispar, 192

magnesium, 128, 135, 137, 179, 246
Magnolia, 231, 232
Magnolia acuminate, 231
maintenance, 135, 163–197, 216
Malacosoma neustria, 192

Malus, 231
manganese, 112, 128, 129, 135, 149, 152, 179–181, 249
market, 17, 24, 34, 53, 67, 80, 82, 153, 155, 216–227
marketing, 15, 211, 212, 215, 219, 220
Mattirolomyces, 52, 55, 82, 83
Mattirolomyces terfezioides (Mattir.) E. Fisch., 52, 82, 83
Mattirolomyces tiffanyae Healy, 83
media, 45, 48, 119
Melanogaster, 90
Melanogaster ambiguous, 90
Melia azedarach, 231
Melicytus ramiflorus, 232
Melissopus latiferreanus, 191
methods and techniques, 24, 34, 35, 44–52, 57, 58, 95–97, 114, 154, 155, 162–166, 199–209, 226–229
Metrosideros, 232
Metrosideros excelsa, 232
microscope, 40, 59, 68, 70, 91, 95, 223
Microsphaera alni, 185
Microsphaera alphitoides, 185
mildew, 185, 186
molecular techniques, 54, 57, 63, 69, 76, 82, 86, 95–97, 108, 115, 155, 226
Morchella, 55
Morel. *See Morchella*
Morus, 232
Musa, 231
Mycoclelandia bulundari, 90
mycorrhiza, arbuscular, 36–38, 51, 124, 126, 133, 139, 140, 231
Myrica, 231
Myzocallis coryli, 191

Nectria cinnabarina, 187
Nectria ditissima, 187
Nectria galligena, 187
nematodes, 38, 44, 177, 190
Nezara viridula, 191
nocciolo, 53, 237
noisetier, 53, 237

nomenclature, 53, 54
nutrient deficiency, 178

Octavianina, 90
odour, 17, 63, 64, 74, 76, 80, 193, 199
Olea europaea, 140, 232
Olearia paniculata, 231
olive oil, 186, 190, 223
Oregon black truffle. *See Leucangium carthusianum*
Oregon spring white truffle. *See Tuber gibbosum*
Oregon winter white truffle. *See Tuber oregonense*
organic carbon, 246
organophosphates, 192
Orgya antiqua, 192
ornamentation (of spores), 59–83
Ostrya carpinifolia, 106, 112, 234, 237
Ostrya virginiana, 83

Pachyphloeus, 55, 88
Pallier system of truffle cultivation, 166
pathogens, 3, 119, 156, 159, 163, 166, 182–189, 193–195
Paulownia, 232
peat, 48, 52, 190
pecan. *See Carya illinoiensis*
pecan truffle. *See Tuber lyoniae*
peridium, 58–88
Périgord, 15, 19–21, 24, 29, 216, 220, 224
Périgord black truffle. *See Tuber melanosporum*
Persea americana, 231
pests, 44, 130, 156, 163, 183, 184, 186, 189–193
Peziza, 55
Peziza arenaria, 55
pH, 48, 52, 70, 98–120, 122, 124–135, 139, 150, 160–165, 175–180, 214, 246
phenol, 64
pheromone, 204
Phormium tenax, 231
phosphorus, 36, 119, 128, 177, 181, 245

Phyllactinia corylea, 185
Phyllactinia guttata, 185
Phylloxera, 35, 124
Phytophthora cinnamomi, 187
Phytophthora infestans, 186
Phytophthora kernoviae, 156, 159, 187
Phytophthora ramorum, 156, 159, 187
Phytoptus avellanae, 191, 192
pianello, 21
Picea, 233, 234, 238
Picea excelsa, 234
Picoa, 55, 85
Picoa carthusiana. *See Leucangium carthusianum*
Picoa juniperi Vittad., 85, 241
Picoa lefebvrei (Pat.) Maire, 85, 241
pigs, 30, 126, 137, 194, 204–206, 209
Pinus, 233
Pinus armandii, 70
Pinus brutia, 234, 238
Pinus halepensis, 118, 234, 238
Pinus nigra ssp. *nigra*, 234, 238
Pinus nigra ssp. *nigricans*, 238
Pinus pinaster, 118
Pinus pinaster var. *atlantica*, 235
Pinus pinea, 97, 117, 118, 235, 238
Pinus radiata, 41, 156, 182
Pinus strobus, 235, 238
Pinus sylvestris, 235, 238
Pinus tabulaeformis var. *yunnanensis*, 70
Pittosporum, 232
Plagianthus betulinus, 232
plant layout, 141–148
plant quality, 155
planting density, 115, 141–148, 195–197, 248
planting, when and how, 131, 134, 137, 161–166, 187, 190–197
Podocarpus, 232
poisonous, 23, 57, 81, 86, 88, 185
pollution, 120, 184, 212
Populus, 112
Populus alba, 112, 233, 235, 238
Populus nigra, 235, 238

Populus tremula, 235, 238
prices, 43, 66, 69, 72, 78, 81, 162, 209, 211–213, 216–228
production. *See* yields
pruning, 48, 157, 165, 166, 184, 187, 188, 194–196, 248
Prunus, 232
Prunus armeniaca, 231
Prunus avium, 231, 233
Prunus cerasus, 233
Prunus dulcis, 231
Prunus padus, 233
Prunus persica, 232
Pseudomonas syringae pv. *avellana*, 188
Pseudotsuga menziesii, 41, 187, 233, 235, 238
Pulvinula constellatio, 95, 152
pure culture of truffles, 51, 52
Pyrus communis, 232
Pyrus pyraster, 233

Quercus, 118, 233
Quercus alba, 83
Quercus cerris, 94, 112, 187, 234, 237
Quercus coccifera, 234, 237
Quercus coccinea, 80
Quercus faginea, 234, 237
Quercus ilex, 106, 143, 146, 169, 187, 234, 237
Quercus macrocarpa, 83
Quercus pallustris, 54
Quercus petraea, 234, 237
Quercus pubescens, 143, 153, 216, 234, 237
Quercus robur, 37, 54, 106, 128, 234, 237
Quercus rotundifolia, 101
quicklime, 135

rainfall, 103, 105, 112, 115, 120, 134, 148, 164, 172, 242, 247
raking, 78, 180
reforestation, 156
rejuvenation of truffières, 195
reticulum, 62–85
Rhizobium, 229

Rhizopogon, 90
Rhizopogon rubescens, 90
Robinia, 52, 82, 231
Robinia pseudoacacia, 52, 82, 234, 237
rock rose. *See Cistus, Fumana, Helianthemum*
Rosellinia necatrix, 187
Rubus fruticosus, 231

Salix, 233
Salix alba, 112, 235, 238
Salix caprea, 235, 238
sand, 45, 52, 82, 105, 115–120, 126, 127, 201, 219
Scleroderma, 122, 123, 152, 168, 175
Sclerogaster, 90
Sedum, 168
Sequoia sempervirens, 232
Sequoiadendron giganteum, 232
shelter. *See* windbreaks
silt, 105, 110, 118, 126, 127
Sirex noctilio, 191, 192
site preparation, 131
site selection, 121, 124–131
smooth black truffle. *See Tuber macrosporum*
sodium, 12, 48, 183, 246
sodium hypochlorite, 48
soil, 17, 20, 37, 40, 99, 104–118, 121–134
soil analysis, 128, 131, 178, 252
soil cultivation, 131, 134–136
soil modification, 134–136
soil moisture, 98, 112, 115, 138, 141, 148, 165, 173–176, 195, 214, 248
soil physical conditions. *See* soil texture
soil preparation, 131
soil sampling, 128–130
soil sterilants, 126
soil testing laboratories, 130, 252
soil texture, 105, 109, 110, 117, 118, 126, 127
Sophora, 232
Sorbus, 232
Sorbus torminalis, 219
Sphaceloma coryli, 187

Sphaerosporella brunnea, 40, 94, 95, 152, 153

Sphaerotheca lanestris, 185

spines, 28, 60–85

Stephensia, 55, 57

Stephensia bombycina (Vittad.) Tul., 88, 89

sterilization, 45, 48, 51

steroid, 204

substrates, 48

Suillia, 203

Suillia fuscicornis, 193, 201

Suillia gigantean, 193, 201

Suillia hispanica, 201

sulphur, 56, 128, 137, 179, 185, 246

summer truffle. *See Tuber aestivum*

Talon's technique, 34, 35, 44, 50

Tanguy system of truffle cultivation, 165

tar, 63, 64

taxon, 53–55

taxonomy, 53–55, 76, 154, 250

Taxus, 232

techniques. *See* methods and techniques

tensiometers, 176

terfez. *See Terfezia arenaria*

Terfezia, 15, 24, 52, 55, 82–85, 118, 119, 234, 236

Terfezia arenaria (Moris) Trappe, 84, 85, 241

Terfezia boudieri Chatin, 85, 118, 241

Terfezia claveryi Chatin, 52, 85, 118, 120, 162, 241

Terfezia leonis. See Terfezia arenaria

Terfezia leptoderma Tul., 85, 118, 241

Terfezia olbiensis Tul., 118

Terfezia pfeilii. See Kalaharituber pfeilii

terfezoid mycorrhiza, 36, 118

Thaumetopoea processionea, 192

Thelephora, 152, 153

Theobroma cacao, 231

thinning, 160, 195–197

thunderstorms, 23, 24, 103, 112, 114, 173, 174, 175

Tilia, 233

Tilia ×europaea, 234, 237

Tilia americana, 80, 83, 234, 237

Tilia cordata, 112, 234, 237

Tilia platyphyllos, 234, 237

tines, 134, 166, 167, 171

Tirmania, 15, 55, 73, 83–86, 118, 119, 234, 236

Tirmania nivea (Desf.) Trappe, 84, 85, 241

Tirmania pinoyi (Chatin) Malençon, 84, 85, 241

Tomentella, 122

Tortrix viridana, 192

toxic. *See* poisonous

trace elements, 128–131, 133, 137, 151, 179–183, 195, 246–249

tractor, 131–133, 135, 166, 172

tradition, 85, 120, 121, 161, 164–166, 176, 199, 212, 221, 224, 229

traditional system of truffle cultivation, 165

training truffle dogs, 206–209

transformed roots, 51

tree arrangement, 141–148

tree layout. *See* tree arrangement

tree protection, 137–141

tree selection. *See* host plants

tree shelters, 192, 253

Tricholoma matsutake, 38, 41, 42, 44, 48, 91, 99–101

truffière, 21, 35, 42, 44, 48, 91, 98–178

truffle beetle. *See Liodes*

truffle diseases, 193

truffle dog, 86, 88, 99, 160, 206–209

truffle fly. *See Suillia*

truffle plantation. *See* truffière

truffle, definition, 55

Tsuga, 233

Tuber aestivum Vittad., 17, 18, 61–63, 92, 94, 106–110, 147, 234, 236, 250

Tuber albidum. See Tuber borchii

Tuber asa Tul. & Tul., 76

Tuber borchii Vittad., 17, 18, 39, 73–76

Tuber brumale var. *moschatum* De ferry, 250
Tuber brumale Vittad., 42, 63, 64, 92, 93, 152, 157, 223, 239, 250
Tuber dryophilum, 75, 76, 77
Tuber excavatum Vittad., 88, 89
Tuber foetidum Vittad., 88
Tuber formosanum Hu, 67
Tuber gibbosum Harkn., 17, 78, 79
Tuber gigantosporum Y. Wang & Z.P. Li, 67
Tuber himalayense. See Tuber indicum
Tuber huidongense Y. Wang, 67
Tuber indicum Cooke & Massee, 17, 67–70, 92, 94, 97, 214, 223, 250
Tuber lyoniae Butters, 80, 81
Tuber macrosporum Vittad., 50, 64, 66, 92, 93, 223, 234, 236, 250
Tuber maculatum Vittad., 18, 73–77, 95, 121, 131, 141, 154, 223
Tuber magnatum Pico, 15, 16–19, 50, 72, 73
Tuber melançonii Donadini, G. Riousset & Chevalier, 88
Tuber melanosporum Vittad., 15–19, 28, 31, 37, 39, 42–60, 91, 94, 99–105, 125, 141–146, 154–156, 167–171, 174, 216–218, 223–227, 234, 236, 239, 240, 250
Tuber mesentericum Vittad., 64, 65, 92, 223, 234, 236, 250
Tuber oligospermum (Tul. & Tul.) Trappe, 73–78
Tuber oregonense nom. prov., 85, 223, 241
Tuber panniferum Tul., 88
Tuber pseudoexcavatum Wang, Moreno, Manjón & G. Riousset, 67, 70, 71
Tuber pseudohimalayens. See Tuber indicum
Tuber puberulum Berk. & Broome, 74, 75, 77
Tuber rapaeodorum Tul. & Tul., 76

Tuber requienii Tul., 88
Tuber rufum Pico, 88, 89
Tuber scruposum R. Hesse, 76
Tuber sinense. See Tuber indicum
Tuber spinoreticulatum Uecker et Birdsell, 70
Tuber texense. See Tuber lyoniae
Tuber uncinatum. See Tuber aestivum
Tuber zhangdianense X.Y. He, H.M. Li & Y. Wang, 67
Tuberaria, 235, 239

Ulex europaeus, 231

veins, 60–88
Verpa, 55, 56
vesicular-arbuscular mycorrhiza. *See* arbuscular mycorrhiza
Viburnum, 232
Vitis, 231

warts, 60–70, 85, 88
weeds, 150, 166–168, 171–173, 248
Weinmannia racemosa, 232
white truffle. *See Tuber borchii, Tuber gibbosum, Tuber magnatum, Tuber oregonense*
wholesalers, 53, 217, 218, 220
windbreaks, 137–139, 192, 253
winter truffle. *See Tuber brumale*

Xanthomonas arboricola pv. *Corylina*, 156
Xanthomonas campestris pv. *Corylina*, 188

yeasts, 56, 229
yields, 14, 120, 142, 157, 168, 174, 200, 211, 215, 216, 248

Zeuzera pyrina, 189
zinc, 128, 129, 135, 179, 181, 183